An Introduction to

DEEPWATER FLOATING DRILLING OPERATIONS

FLOATING DRILLING EXPERIENCE

Water Depth Records, Worldwide

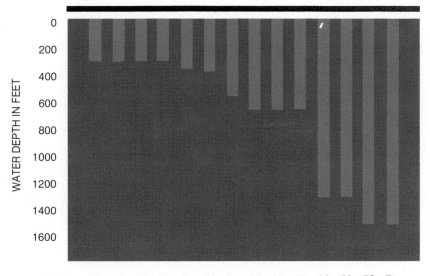

An Introduction to

DEEPWATER
FLOATING
DRILLING
OPERATIONS

L. M. Harris

PETROLEUM PUBLISHING COMPANY
Tulsa

Library of Congress Catalog Card Number: 72-76603
International Standard Book Number: 0-87814-011-5
Printed in U.S.A.

Acknowledgments

In developing the information presented in this text, the author acknowledges the cooperation and assistance of many individuals and organizations. Photographs, drawings and data contained herein have been made available by

American Bureau of Shipping
Society of Petroleum Engineers of AIME
American Petroleum Institute
American Society of Mechanical Engineers
Baldt Anchor and Chain Co.
Bethlehem Steel Corp.
Cameron Iron Works
Fluor Drilling Services, Inc./Western Offshore Division
Global Marine Inc.
Offshore Technology Conference
Otis Engineering Corp.
Sedco, Inc.
C Jim Stewart & Stevenson, Inc./Oilfield Division
United States Coast Guard
Vetco Offshore Industries, Inc.

In addition the author appreciates the previous work experience with Humble Oil and Refining Company and its training in the standards necessary for full reliability and safety in offshore operations. And most particularly, he is indebted to his wife, Mary, and her untiring efforts in typing and retyping the manuscript.

L. M. Harris

Contents

Preface

Prior to 1950, the petroleum industry limited its offshore drilling activities to those conducted from bottom-founded structures. True, some floating vessels were used in punch-coring operations, but it was not until the 1950's that rotary-drilling operations were first conducted from floating vessels. Even then, most of the activity was limited to the coring type of operations. That is, the production or final casing string was not set and no attempts were made to surface hydrocarbons with a drill-stem or production test from a floater.

Most of the development of equipment and technology for rotary drilling from floating vessels started with an extensive shallow-hole coring operation off the coast of California in the Santa Barbara Channel. The very first operations were conducted "over the side" of the drill vessel, but these were quickly supplanted by vessels designed with a center drill well or "moonpool".

These early efforts utilized subsea wellhead and blowout-preventer stacks mounted on the ocean floor very much like those used today, but it was not until the late 1950's that the marine-riser concept was born. The first units utilized a rotating head that latched into the subsea stack, and mud returns were taken through a hose connected to a side outlet on the stack. The blowout preventers were conventional land-rig preventers without modification; they were operated with direct-function control hoses, bundled and extending between the drill vessel and the subsea stack.

Development of equipment and technology accelerated

during the 1960's. This was the period that the semisubmersi-bles, the pilot-operated control system, the tensioned-riser concept, the acoustic position-reference system, and subsea completions all came into being.

The 1950's saw drilling in the 200 to 300-ft water- depth range. The 1960's have now seen drilling in 1,000 to 1,300 ft of water. The 1970's will surely see drilling commonly in 2,500 to 3,000 ft of water and probably some activity in 5,000 ft of water.[1]

The 1970's will certainly see subsea completions perfected and used extensively. To date these completions have been limited to prototypes, to the shallow water, and usually to diver-assisted design.

This text has been prepared, however, to discuss what can be done today—the technology and the equipment, its capabili-ties and its limitations, and how it can be used safely to accomplish the job at hand.

L. M. Harris
April, 1972

1. The author acknowledges the fine accomplishments of Global Marine and the Scripps Institute of Oceanography with the Challenger in much deeper water depths. Here, however, he refers to those drilling operations of a continuous nature that permit controlled communication with the well bore and extended drilling operations, as opposed to the "single-trip" coring opera-tions of the Challenger.

An Introduction to
DEEPWATER FLOATING DRILLING OPERATIONS

1

Introduction   ✳
to Floating Drilling
Operations

In less than 20 years, the science of drilling from floating
vessels has developed basically from the scooping of a sample
of formation from the ocean floor. Now it has reached the
highly complex technology involved in drilling in water depths
of up to 1500 ft, with all the safety, control, and capability
available to land-based operations. The equipment and techniques are less widely known than conventional land drilling,
the investments in equipment and personnel are much
greater, and good solid planning and supervision are even more
essential to efficient, economical ventures.

The floating drilling system is composed of four major
components, or subsystems, that should be identified to simplify terminology and to help in our understanding of the
overall floating system. These are:

1. The drill vessel.
2. The mooring system.
3. The well-control and communications system.
4. The drilling system.

Each of these has many further subdivisions that will be
discussed in detail. Then, too, there are the essential auxiliary
or support facilities that are involved in floating drilling

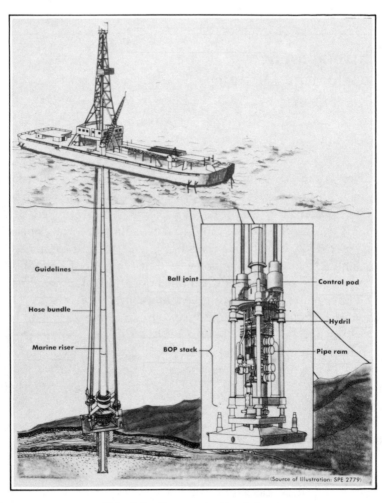

FIG. 1.1 Floating Drilling System

2

operations. These include the shore facilities, auxiliary vessels, aircraft, communications and data gathering and monitoring equipment. All of these systems, subsystems, components, and support and auxiliary facilities need to be fully understood and integrated for the design of a safe, efficient floating drilling operation.

Three types of floating drilling vessels are in common use today—the semisubmersible, the ship, and the barge.

The semisubmersible has the primary design advantage of placing the bouyancy chambers below the surface of the water. This minimizes the response to the wave forces, hence stabilizing the motion of the working floor and drill string.

The ship type has the conventional ship-form hull and has the chief advantage of being able to travel from location to location at reasonably good speeds by its own power.

The barge type commonly has the same hull form as the ship type but is not self-powered.

Advantages and disadvantages of each type for particular operations will be developed later.

Some current mooring systems utilize the conventional straight chain system or a pure wireline system, but most are a combination system of chain and wire. Water depth and oceanographic conditions are generally the controlling factors in selecting the mooring system for the vessel.

Several vessels have been built, or are in the design and construction stage, that depend entirely upon thrusters, or powered units, to position the vessel dynamically. Best known of these is the Global Marine Challenger that has been active in the extremely deep water (5,000 to 20,000 ft) on single-trip coring-type operations.

The well-control and communications system can be considered the heart of floating drilling. This system provides for the control of, and communication with, the well bore during drilling operations. It includes such components as the casing, the wellhead, the subsea blowout-preventer stack, the flexible joint, the marine riser, the kill and choke lines, the telescoping joint, the tensioning system, and all of the controls, guidelines, and tools necessary for the installation and operation of the system components.

We refer to the drilling system as that system that includes those components normally found in conventional land drilling operations, such as, the mast or derrick, the rotary, the drawworks, the pumps, the drill pipe and collars, the drill tools, and the power equipment. The only major difference in the drilling system on the floaters is the use of motion-compensation equipment, generally bumper subs located deep in the drill string.

2

The Unique Aspects of Drilling From a Floating Vessel

Many conditions in drilling from floating vessels differ from those in normal land operations or even in drilling from fixed offshore structures. Three areas of major differences are readily identifiable:

1. Motion and motion compensation
2. Remote location of well-control equipment
3. Weather and oceanographic influence on operations

The design of a floating drilling system includes an analysis of these differences to determine their possible consequences and to establish methods to insure full reliability. This analysis becomes the basis for equipment selection and practices for floating drilling operations. Additional practices common to all land drilling operations but critical to well control and blowout prevention are then included to provide safety.[1]

Vertical motion and its compensation

The dominant effect of the vertical component of the motion of the drill vessel is the relative motion between the drill bit and the well bore. Motion compensation is required to keep the bit on bottom and to control bit weight. Slip joints (bumper subs) located in the upper portion of the drill-collar section are commonly used for this purpose. An inherent extra cost of this technique arises from the inability to vary bit weight without pulling the drilling assembly. In addition, extra trips and rig time are often required to service the slip joints. A strict repair and replacement schedule can reduce this lost time.

Slip joints compensate for vertical motion only in the lower section of the drill string. The upper drill pipe is constantly moving vertically through the marine riser, the blowout preventers, the wellhead, and the casing strings. Wear is minimized in these areas by eliminating the use of hardbanded tool joints. Furthermore, when any of the blowout preventers is closed on the drill pipe, wear on the preventer is prevented by hanging a tool joint on one of the rams.

The industry has had only limited experience with other means of vertical motion compensation. Humble Oil & Refining Co.'s SM-1, a small vessel used in the Santa Barbara Channel in the 1950's for coreholes, utilized a modified torque converter that provided slippage for the drawworks. The system worked efficiently for the loads encountered in shallow wells.

In another approach to surface motion compensation, a pneumatic tensioning system has been recently installed on a vessel on the U.S. West Coast. Aim is to control drilling-line tension. Several manufacturers are designing and testing other devices to provide compensation aboard the drill vessel. Surface compensation has real promise of significantly reducing costs in floating drilling operations.

For logging or other wireline operations, vertical motion

compensation has been handled adequately with surface sheave arrangements and pneumatic tensioning cylinders.

Horizontal vessel displacement

All mooring systems are designed to resist environmental conditions. For floating drilling vessels, mooring systems must not only provide satisfactory restoring force, but also limit horizontal displacement. In the Santa Barbara Channel, Humble had designed moorings to limit this displacement to 10% of water depth. In practice, however, displacement has not exceeded 3%.[2]

Consequences of horizontal movements are:

1. Fatigue stressing components of the drilling system, particularly the marine riser system.
2. Increased wear of the subsea components created by misalignment.
3. The requirement for vessel position monitoring and control.
4. If excessive, the necessity of disconnecting until the horizontal motion is limited.

Inadequate control increases downtime and causes premature replacement for marine risers, ball joints, blowout preventers, and similar equipment.

Reliable information on the relative position of the drilling vessel and the subsea wellhead is a key to limiting horizontal displacement. Two types of monitoring devices are in use today. Most common is the taut-line system. The system consists of:

1. A taut steel line stretched from the vessel to an anchoring point on the ocean floor. It may be a guideline or an independent line with an attached anchor weight.
2. A dual-axis inclinometer to measure the slope of the line.

The taut line is assumed to be straight from the vessel to the ocean floor, so its slope indicates vessel displacement in the horizontal plane. It has been shown that, as water depth increases, drag forces due to current can distort the taut line appreciably, so the assumption of a straight line may be invalid. The position indication given by the system could then be quite inaccurate.[3]

Another system is the acoustic position-reference system. Basic components of this system are:

1. An acoustic transmitting beacon located on the ocean floor.
2. Hydrophone receivers mounted below the vessel.
3. A vertical sensing unit to compensate for vessel motion.
4. A computer control and display console.

Using the acoustic signals, the display console shows continuously the location of the drilling vessel as to magnitude and direction of the departure from the wellhead. Units are available with accuracy within 0.5% of water depth.

Remote location of well-control equipment

The location of well-control equipment on the ocean floor introduces complexities not known in land or fixed-platform drilling. Technology and equipment are available to offset these apparent disadvantages. The normal monitoring equipment used on land rigs (flow indicators, pit level devices, hydraulically operated flow chokes, etc.) are all useful for the floating rig also. In addition, hydraulic, acoustic and electric monitors and controls provide the extra capability to counterbalance the effects of the remote installation. Well designed and carefully selected equipment and procedures, backed up with redundancy, have been proved in water depths exceeding diver capability.[4]

Weather and oceanographic influences

In any offshore operation, the problems of supplying and transporting material and personnel are directly related to weather and oceanographic conditions. For the floating drilling operations, the drill vessel itself and all drilling operations are also affected by the environmental conditions. The particular effects of vertical and horizontal motion have been discussed.

A combined motion effect, such as from heave, pitch, and roll in varying combinations, have different effects on the different operations. In general, all motion affects all operations and increases costs. The simpler operations, such as routine drilling and tripping of the drill string, are less affected by the environment. The more complex operations, such as running and landing of the subsea blowout-preventer assembly, are more restricted by environmental conditions. Hence, they are more costly during severe vessel motions.

A complete analysis of all of these aspects of drilling from a floating vessel together with the possible consequences and problems and methods to compensate or reduce risks is presented in Table I.

TABLE I

Problems Unique to Drilling from a Floating Vessel

Problem	Consequences	Corrective Measure
Vertical Motion and Motion Compensation		
1. Abnormal wear between drill pipe and the preventers, wellhead, and casing.	1. Loss of pressure integrity of subsea or subsurface equipment.	1. Prohibit use of hard-banded drill pipe.
2. Compensation for motion with bumper subs introduces a weakness in the drill string from structural standpoint and for pressure integrity (seals).	2. a. Leaks develop in the drill string necessitating tripping the bit prematurely; b. The drill string separates at the bumper sub necessitating a fishing job; or c. Either of the above occurring during a well-control operation increases mud density required for well control.	2. a. Schedule the repair and replacement of bumper subs more frequently; b. evaluate and select bumper subs from manufacturers with best performance records; or c. eliminate bumper subs with vessel motion compensation.
3. Severe wear on the sealing elements of the preventers resulting from vessel motion while the preventers are closed. Tool joints if improperly located could cause extreme loading on closed pipe rams.	3. a. Loss of sealing capability of the preventers; or b. in the case of tool joints just below pipe rams, the drilling string could be parted or the drilling mast collapsed.	3. a. Exercise extreme care in use of pipe rams to consider location of the tool joints; b. land the pipe on the rams as soon as practical; c. include a float valve in the drill string to permit disconnect on the drill vessel; d. provide vessel motion compensation.
4. Modifications of normal techniques of logging, varying bit weight, directional drilling, drill-stem testing.	4. Increased rig time and requirements for extra rig and well equipment.	4. Vessel motion compensation.
Horizontal Vessel Displacement		
1. Requires vessel-position monitoring and control.	1. Increases rig investment.	1. None

2. Increased wear and fatigue stressing from any misalignment.

3. Possibly require disconnect, or cause shear failure, either creating a loss of communication with well bore and with control equipment.

Remote Location of Well-Control Equipment

1. Remote monitoring of well conditions.

2. Remote control of preventers, choke and kill valves.

3. More difficult maintenance and inspection of well-control equipment.

Influence of Weather and Oceanographic Conditions on Operations

1. Increased downtime.

2. Sudden and significant changes in weather will cause corresponding changes in vertical and/or horizontal vessel motion.

2. Leak or structure failure in riser, preventers, or wellhead equipment.

3. The loss of a well bore or even the loss of control of a well.

1. Delayed knowledge of conditions at wellhead.

2. a. delayed reaction time; and, b, possible hydrocarbons above control point.

3. a. time delay for repair and inspection; and

   b. limited capability of repair men.

1. Increased costs.

2. All problems associated with vessel motion.

2. a. Increase accuracy and reliability of monitoring equipment;
   b. strengthen mooring system.

3. a. Increase accuracy and reliability of monitoring equipment;
   b. strengthen mooring system;
   c. provide float valve in the drill string and provide automatic or acoustic control on selected preventers.

1. Acoustical/electrical monitors.

2. a. acoustical/electrical controls;
   b. increased hose sizes for hydraulic systems; and,
   c. diverter on marine riser.

3. a. rigid selection of subsea equipment and standard operating procedures; and
   b. additional equipment to provide full redundancy.

1. a. improved design and selection of hull configuration;
   b. strengthened mooring system; and,
   c. additional weather forecasting.

REFERENCES

1. Harris, L. M.: "Design for Reliability in Floating Drilling Operations." Second Offshore Technology Conference, OTC 1157, April 1970.

2. Harris, L. M., and Ilfrey, W. T.: "Drilling in 1300 Feet of Water—Santa Barbara Channel, California." First Offshore Technology Conference, OTC-1018, May 1969.

3. Adams, R. B.: "Accuracy of the Taut-Line Position Indicator for Offshore Drilling Vessels." 1967 Petroleum-Mechanical Engineering Conference, ASME, Paper No. 67-Pet-5.

4. Harris, L. M.: "Floating Drilling Experience in Santa Barbara Channel." 1969 California Regional Meeting of the Society of Petroleum Engineers of AIME, SPE-2779.

3

Planning and Organizing for a Floating Drilling Operation

As soon as management decides an offshore prospect has merit and to risk capital in the venture, staffing becomes the first order of business. The number of people to be involved varies as to timing, experience level of the people selected, size and type of operation, and operating policies established by management.

Just as in land drilling operations, certain basic parameters need to be analyzed first. These include the proposed spud date, the number and type of wells to be drilled, the number of rigs to be used, and the availability of rigs.

For floating drilling, new parameters must also be considered. First are the weather and oceanographic conditions for the specific location and time of year. Sea conditions in most areas do vary significantly through the year. Other considerations are the availability of service facilities, including boat docks, shore bases, helicopter ports and service contractors, and the proximity of population centers and other marine acitivities.

Water depth has important influence on the planning, rig and equipment, and personnel needed for the operation.

ANALYSIS OF DRILL VESSEL AND EQUIPMENT

Most companies today utilize contractors' rigs and personnel for floating drilling operations. The discussion here will be directed to that assumption. Even if company-owned rigs are employed, much the same analysis will be required of that equipment.

Drilling contractors should be supplied as much information as possible on the proposed venture. They certainly need information on the number and type of wells to be drilled, the proposed start-up date, the location, and the water depth. From this, they can offer proposals for certain rigs that will be suitable and available for the operation. Generally, they will also offer counterproposals, particularly in regard to start-up time. If a company has some flexibility, it can often get a much better vessel at less overall cost.

Mobilization often becomes a major part of the cost of the venture. Certainly this is true for the short-duration (one- to three-well) programs.

In evaluating the contractors' equipment, consideration must be given not only to the rig proposed and its suitability for the job. Also the possibility must be considered that: (1) the original well designs might be modified to fit the available vessel; or (2) the vessel might be modified to fit the job.

For instance, the well may be relocated in shallower water and directionally drilled if the vessel is limited on water-depth capability. Or, the hole size might be reduced to permit the use of a smaller rig. Changes in plans can be economically attractive for the short duration programs with their potentially high mobilization costs.

On the other hand, for a sufficiently long program, investments in rig changes and the addition of extra equipment can usually be justified as a means to improve efficiency. Certainly, where high mobilization costs are necessary to move in the exact rig needed, consideration should be given to spending money on rigs locally available to bring them up to specifications.

Sacrifices in efficiency and changes in well designs are often justified. Sacrifices in safety or the lowering of standards for reliability should never be accepted.

Drilling equipment

Much of the equipment on the drill vessel is evaluated by the same standards used for land rigs simply because the equipment is used to perform the same basic functions. This includes the mast, the drawworks, the mud pumps and mud tanks, drill collars, drill pipe, instrumentation, and cementing and logging units.

One word of caution on the drill pipe—drill pipe with hardbanding of any type should never be used on the floating drilling vessel. Motion compensation is provided in the lower part of the drill string, but the upper part moves continuously with the heave of the vessel. And it continually moves the drill pipe vertically in relation to the marine riser, the blowout-preventer stack, the wellhead, and the casing.

The price of drill pipe must be weighed against the cost of floating drilling operations and the potential danger and cost of subsea wear. One cannot justify transferring this wear from the drill pipe.

Then, too, in a case where the well has to be closed in, and the drill pipe run deeper in the hole, tool joints with hardbanding cause additional wear on the seal elements of the preventers.

BOP stack and controls

Many contract drilling vessels are equipped with, and commonly use, a "two-stack" blowout-preventer system. One stack is usually of large diameter (20¾-in.), low-pressure (2,000-psi) and is used for drilling and setting casing in the shallow first part of the hole.

After setting these casings, the stack is removed and a smaller (13⅝ or 16¾-in.) and higher pressure (5,000-psi) stack is run for the remainder of the drilling and completion operations.

The first stack generally consists only of a Hydril-type preventer and, maybe, one ram-type preventer. The second, higher-pressure stack consists of one or two Hydril-type preventers and three or four ram-type preventers.

This change-out of preventer stacks can easily cost two days of rig time. Hence, well designers should give serious consideration to casing designs that avoid the need for the first, large-diameter stack. Some companies use slim-hole designs and special-clearance couplings on casings. Through these, they manage to design 12,000-ft wells with three casing strings and use a single 13⅝-in stack system. The single 16¾-in. stack, of course, offers more flexibility than the 13⅝-in.

An integral part of the evaluation of the blowout-preventer system is the evaluation of the control system to operate the preventers. Particularly to be considered are the time delays built into the system by the nature of its design and equipment. Two general types of hydraulic control systems are commonly used today. The first, the direct system, is simply an extension of the systems used on land with longer control lines. In it, hydraulic fluid is directed to the individual preventers, kill and choke valves, and other hydraulically operated equipment on the subsea stack by simple direct lines or hoses. The flow is directed through valving on board the drill vessel. The system is good, simple, and entirely adequate for operations in shallow water.

In deeper water, however, line friction and hose expansion add significant delays to the reaction time (the time from first operating a control valve at the surface until the function is fully performed on the ocean floor).

The indirect control system offers the advantage of reducing both the reaction time and the size of the hose bundle between the drill vessel and the ocean floor. In the indirect control system, pilot valves are located on the subsea stack and are actuated from the drill vessel hydraulically. A common power-fluid supply is then directed on the stack to the desired function. This supply comes from accumulators on the drill vessel down a common power-fluid line to the subsea stack. Sometimes, it is supplemented from accumulators mounted directly on the subsea stack. The type and capacity of these accumulators, both surface and subsea, affects the reaction time of the blowout-preventer system.

Marine riser

The amount and type of marine riser and the riser-tensioning equipment available from the contractor are important in evaluating the rig as to water-depth capability. Not only should consideration be given to the amount of riser needed to reach between the ocean floor and the drill vessel but also to providing back-up or insurance against the unexpected loss of a riser. The accidental dropping of a riser, or damage from any source, can cause a long and expensive delay in waiting on a replacement.

Inside diameter of the riser pipe and its size compatibility with the subsea blowout preventers and the proposed well program are important to the potential customer. The telescoping joint, its diameter and length of stroke, and the type of flexible joint, its diameter and maximum angle, need consideration. The riser-tensioning system and the guideline system should be analyzed as to suitability and capability. The tensioning requirements on both these systems increases with water depth and water currents. These requirements are also influenced by mud weight, wave and current forces, and vessel displacement.

Mooring system

All reliability and security designed and built into the various components of the floating drilling system depend upon the single reliable functioning of the mooring system. All other systems are designed on the assumption that the vessel can be held on the location. Critical analysis of the mooring components is essential.

Chain, wire, and combinations of chain and wire are in common use for mooring. All can be satisfactorily used if designed for the oceanographic forces to be encountered by the drill vessel. As a general rule, all of the links in the system should be designed with a factor of safety of three when in new condition. This gives some leeway for chain or line damage, wear, and fatigue stressing.

Size and type of anchors needed varies with water depth, length of mooring lines, and ocean-floor conditions. When installed, each mooring leg should be tested to prove that the anchors have sufficient holding power to withstand the maximum loads expected.

Remote readout and remote control of anchor winches are

highly desirable but not essential to safe operation. Vessels so equipped, however, are easier to control during inclement weather.

Quarters, storage capacity

The size of accommodations needed for personnel and the storage capacity needed for tubular goods and mud products depend on the location, proximity to suitable land facilities, and the environment. Most floating-vessel ventures require that personnel be housed on board the vessel. They can be accommodated on shore, however, if it is close enough and if vessel space is at a premium. The same is true for storage of supplies and well equipment. Weather does have its influence, in that rough seas or poor visibility can delay transportation and add to cost.

Hospital facilities, bath and toilet facilities, the galley, and recreation facilities all should be considered in rig selection. Deck-load limitations frequently are important to the remote-location operations.

Safety, emergency facilities

All points discussed up to this point have a great bearing on safety—safety for personnel, for the environment, for the operation, and for the equipment. One of the best measures of safety, however, is the examination of past experience. Records usually are available to the prospective customer. Assurance that station bills are posted and that fire-control and life-preserving equipment and drills meet regulatory standards is usually sufficient to assure adequacy. The provision of equipment and practices for controlling pollution is becoming more and more important to the operating companies and to the industry in general.

General condition of equipment

All paper analysis and design specifications cannot supplant a personal inspection of the physical condition of the various components and of the general appearance of the equipment offered. Evidence should also be examined on the last shipyard inspections of the vessel, the mooring equipment, and the components of the subsea equipment. The criteria for acceptance of this information depend, in part, on the critical nature of the operation to be undertaken and the original factors of safety in the design.

Personnel

The method of evaluating personnel offered by a contractor differs little whether the job be a floating drilling operation, a land drilling operation, or the construction of a private residence. People should be evaluated on the basis of their number, their experience, and their past performance. The equipment is certainly no better than the people available to operate it.

Support equipment and service

Certain support services are required for all offshore ventures. These include auxiliary vessels (crew boats, supply boats and tugboats), helicopters, and shore facilities. On board the vessel, additional equipment is needed—communications, subsea television, navigation, and position-monitoring equipment. Some of these facilities may be offered by the drilling-vessel contractor and should be considered in the light of the requirements for the operation.

Checklists

A checklist entitled "Preliminary Planning and Rig Evaluation Checklist" has been prepared and is included as Appendix A. It summarizes the material covered in this chapter and presents it in readily usable form. It is designed to evaluate the capability of the vessel and its equipment to do the job.

A companion checklist prepared by the U.S. Coast Guard provides guidance in establishing the seaworthiness of the vessel and will be presented in a later chapter.

ANALYSIS OF PERSONNEL REQUIREMENTS

In determining the number of personnel required to staff the offshore floating drilling venture, two general categories of people need to be considered, i.e.: those required *on board* the vessel and concerned directly with the operation; plus those required *on shore* who are primarily a management-technical support group. Company and contract personnel will probably be involved in both groups.

On board

Company personnel on board the vessel include both supervisory and technical personnel. A company drilling supervisor should always be on board and in charge of, and responsible for, the entire operation concerning the vessel. All contract personnel on the vessel should report through their supervisors to this one man. He should be on duty 24 hours a day and not have a tour-type assignment. Relief will have to be provided for days off, of course, and the schedule of this relief depends on the duration of the assignment and the type of the operation.

The drilling supervisor should be responsible to, and report directly to, the district superintendent located at the shore base. The drilling supervisor, being in charge of the operation, will: (1) direct contract personnel; (2) set and maintain standards of safety and performance; (3) control transportation and communications to and from the drill vessel; and (4) anticipate and procure supplies, equipment, and services.

Company technical personnel, including one drilling engineer and one wellsite geologist, should be on board the vessel most of the time. The drilling engineer should: (1) evaluate and recommend drilling parameters and equipment and material-handling practices; (2) evaluate equipment performance; (3) recommend equipment inspection, maintenance and replacement; (4) accumulate basic oceanographic, equipment, and well data; and (5) provide technical assistance to the drilling supervisor.

The wellsite geologist should: (1) direct the collection and the dissemination of geological data; (2) forecast geological horizons; and (3) provide technical assistance to the drilling supervisor.

Contract personnel are generally set up into two separate functional groups, i.e.: the marine group under the direction of the ship's captain (or the barge master); and the drilling and roustabout group under the supervision of the contract tool pusher. The nature of the work of the marine group is such that it can usually be handled under the supervision of one man on duty 24 hours per day. Considering the magnitude of the work to be directed by the tool pusher and the high cost of floating drilling operations, this work is best directed by men on tour duty, preferable two men on 12-hour shifts alternating the responsibility.

On shore

The total offshore venture should be under the direction

of a company manager. He should be responsible for: (1) coordinating the activity to accomplish company goals; (2) establishing operating policies and work objectives of the group; and (3) directing government and public relations.

Reporting directly to the company manager should be a district superintendent. His duties should be to: (1) direct the drilling operations; (2) assist the manager in the performance of his duties; and (3) act for the manager in his absence. As mentioned, the drilling supervisors on board the vessel should report to the district superintendent.

The technical staff should include both engineers and geologists. The district engineer's duties should include: (1) organizing and directing the total engineering effort; (2) establishing goals and objectives in engineering; and (3) assisting management in setting design and operating policies. His staff would: (1) evaluate and recommend drill-vessel selection; (2) design wells and well programs, (3) provide surveillance of the drilling operation; (4) monitor the performance of the drill-vessel equipment including the mooring system and the well-control and communications system; and (5) provide assistance in cost analysis and cost forecasts.

The district geologist would be responsible for: (1) organizing and directing the geological effort; (2) coordinating with management and engineering on geological objectives; (3) assisting management and engineering on geological interpretation during drilling operations; and (4) analyzing and reporting geological data.

Company administrative personnel under the direction of a district administrator would be responsible for: (1) budget forecasting; (2) cost accounting; (3) regulatory permits and reporting; (4) material procurement, inventory and surplus disposal; (5) material expediting; (6) personnel administration; and (7) office, building, and transportation administration.

Contract personnel reporting directly to company personnel would ordinarily be limited to those working under the administrative group in the handling and maintenance of material and equipment on shore.

Contract services personnel, even though based on shore, would generally be under the direction of the drilling supervisor in that their services would normally be utilized only on the drill vessel.

Total company personnel requirements

Total company staff requirements may then be summarized:

Manager: 1

District superintendent: 1

Drilling supervisor: 2 per rig + 1

Engineers: 3 per rig + 1

Geologists: 3 per rig + 1

Administrative: 3 per rig

These are only general guidelines and vary depending on the particular operation and the experience of the personnel.

ANALYSIS OF COSTS AND COST FORECASTING

Just as in a land exploration-drilling or development-drilling program, three general classifications of expenditures are incurred: (1) *venture costs,* or those associated with the mobilization and demobilization of the operation regardless of the number of wells drilled; (2) *well costs,* or those associated with each well regardless of the time spent in drilling it; and

(3) *daily costs,* those associated purely with the length of time that the operation continues. The costs incurred in most floating drilling ventures have been classified by these three categories in Table 2.

Once estimates of each of these costs have been assembled in this form, they can be easily grouped to estimate costs on any alternate plans.

TABLE 2
Cost Analysis Factors

Venture Costs	*Well Costs*	*Daily Costs*
Rig transportation	Site preparation	Rig rental
Rig modifications	Tugboat services	Insurance
Preparation of:	Well equipment:	Fuel and water
Shore facilities	Wellhead	Service boats:
Office	Casing	Crew
Warehouse	Well supplies:	Supply
Docks	Cement	Helicopters
Heliport	Packers	Tool rental:
Mobilization	Well services	Bits
Personnel	Logging	Drill pipe
Services	Perforating	Special tools
Demobilization	Testing	Television
	Fishing	Radio
	Cementing	Services:
		Mud loggers
		Transportation
		Divers
		Roustabout labor
		Drilling fluid
		Rig supplies, repairs
		Company overhead:
		Personnel
		Office
		Hdqrs. overhead
		Miscellaneous

4

The Drill Vessel

✳

Floating drilling rigs are normally referred to as "drill vessels" or "drilling vessels" and are in the truest sense of the word "vessels". For a seafaring man, however, they differ in many respects from his general conception of a vessel.

Traditional vessels that plod the sea carry cargo or passengers from one place to another over fixed routes and on fixed schedules of departure and arrival. The drill vessel does not fit that pattern. It often anchors for weeks and even months at one location. It may move only short distances from well to well. Seldom, if ever, does it need to visit port except for major repair or inspection. Regular vessels are required periodically to go into dry dock for exterior hull inspection. Many drill vessels, semi-submersibles for instance, cannot even be handled in present dry-dock facilities because of their size.[1]

The work environment of the drill vessel differs from that of normal passenger and cargo vessels, too. It has the same exposure to the elements, the sea and its storms, but seldom has the flexibility of the traditional vessels in dodging storm centers. On top of this, add all the usual hazards attendant to any drilling operations in the possible exposure to explosive mixtures of oil or gas.

The U. S. Coast Guard in its analysis of drill vessels, and other special purpose vessels, has established categories of "industrial vessel" and "industrial personnel" for regulatory purposes:[2]

Vessel. "Means every vessel which by means of its special outfit, purpose, design, or function engages in certain industrial ventures. Included in this classification are such vessels as drill rigs, missile range ships, dredges, cable layers, derrick barges, pipe lay barges, construction and wrecking barges. Excluded from this classification are vessels carrying freight for hire or engaged in oceanography, limnology, or the fishing industry."

Personnel. "Means every person carried on board an industrial vessel for the sole purpose of carrying out the industrial business or functions of the industrial vessel. Examples of industrial personnel includes tradesmen, such as mechanics, plumbers, electricians and welders; laborers, such as wreckers and construction workers; and other persons, such as supervisors, engineers, technicians, drilling personnel, and divers."

Industrial personnel cannot be considered either passengers or seamen. Passengers generally know little or nothing of the sea environment. They may even be physically handicapped or infirm and with few exceptions are incapable of self-preservation. Therefore, these people require a high degree of safety protection. The seamen, on the other hand, are able-bodied, trained men with a good knowledge of the marine environment. They have a high degree of capability for self-protection. Industrial personnel fall into a classification between the two. Normally, they are able-bodied and have some knowledge and training in marine environment. They need some intermediate degree of protection.[1]

Thus, though it is common and correct to refer to the floating drilling rig as a "vessel", it was quite different from the traditional passenger and cargo vessel in the minds of most regulatory bodies when most of the current maritime rules were established. The U. S. Coast Guard is acutely aware of these differences. It is working closely with the American Bureau of Shipping and the petroleum industry to develop standards and regulations that are realistic and yet maintain a high degree of safety and protection for the personnel and equipment in floating drilling.

Vessel motions defined

Vessel motions are best described in relation to three axes of the vessel, i.e.: the x-axis extending longitudinally fore and aft; the y-axis running crosswise from the port side to the starboard side; and the z-axis vertical. All three axes have a common point at the center of gravity of the vessel. There are two types of motion in relation to each axis—translational motion and rotational motion. Hence, a total of six motions are recognized: [3]

1. Surge, the translational motion along the x-axis
2. Sway, the translational motion along the y-axis
3. Heave, the translational motion along the z-axis
4. Roll, the rotational motion about the x-axis
5. Pitch, the rotational motion about the y-axis
6. Yaw, the rotational motion about the z-axis

The adaptability of any vessel to service as a drill vessel depends to a large degree on its adaptability to sea conditions and the resulting vessel motions. To minimize the motions, study must be given to typical sea conditions and to the selection of vessel dimensions. Vessel motion, however, is only one factor in vessel selection and must be weighed against all others for safe, economical operations.

Types of floating drilling vessels

Offshore units used by the petroleum industry to explore, develop, and produce oil and gas can be categorized as follows:

1. Fixed platforms, or those units permanently attached to the ocean floor and, for all practical purposes, not moveable.

2. Mobile platforms, or those units that are designed for mobility and intended to be at a given location a limited time period. The three general types of mobile platforms are: (a) the self-elevation units; (b) the surface-type units (ship units and barge units); and (c) column-stabilized or semisubmersible units.

As this text is limited to floating drilling vessels it covers only the two surface-type mobile units; i.e., the ship and barge units, and the column-stabilized or semisubmersible units.

The ship units

Self-propelled drill vessels constructed along the lines of traditional cargo or passenger vessels are known as ship units. Some of these were designed and built for the specific purpose of serving as a drill vessel. Some are simply a traditional vessel converted for this service.

Being self-propelled, they are capable of traveling from location to location without outside assistance. Unlike traditional vessels, they are modified by a center opening through the vessel for conducting drilling operations. This center opening is commonly called a "moonpool". The drill vessels

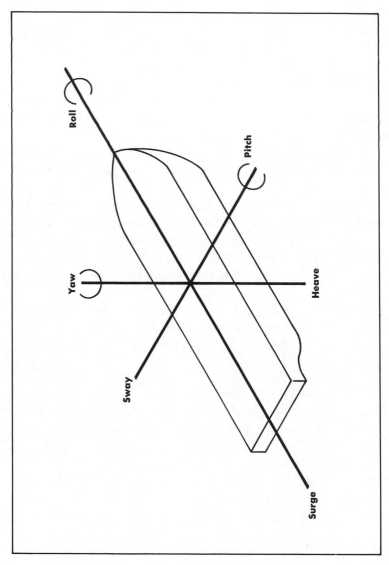

FIG. 4.1 Types of Vessel Motion Defined

30

are also outfitted with a drilling mast or derrick and other equipment necessary for drilling.

Most ships units have drilling displacements of 5,000 to 11,000 long tons. Some, however, have been constructed as small as 450 long tons or as large as 36,000 long tons. There are currently 29 units of this type in service, as listed in Table 3.

The barge units.

Barge units differ from ship units only in that they are not self-propelled and must be towed from location to location. They, too, are frequently converted traditional barges or ships but may be designed and constructed for this specific service. There are currently 24 barge units in use, as listed in Table 4. They vary in drilling displacement from 900 to 10,600 long tons but most are in the range of 3,500 to 7,500 long tons.

The semisubmersible units

The semisubmersible, or column-stabilized, units differ radically in appearance from traditional vessels. They have a platform or deck area that might be square, rectangular, or triangular in general configuration. This platform is supported by columns connected to large underwater displacement hulls, or is mounted on large vertical cassions, or is supported by some combination of the two. Basic purpose of the general design is to reduce wave forces by locating the major bouyancy members beneath the surface or beneath the wave action. In terms of drilling displacements these types of units are much larger than most ship or barge units and range from 8,000 to 17,600 long tons. The 27 of these units in service are listed in Table 5.

FIG. 4.2 Ship Unit: Global Marine, Inc., *Challenger*

FIG. 4.3   Barge Unit: Fluor Drilling Services, Inc., *Wodeco VII*

FIG. 4.4 Semisubmersible Unit: Sedco, Inc., *Sedco 135E*

Types compared

Exploration and drilling operations from floating drill vessels are profit-motivated ventures. Therefore, all evaluations or comparisons of equipment and methods must be based on economics so long as safety standards for personnel and environment are maintained.

In all three types of floating vessels, initial investment is the major factor in determining daily operating costs and, to some degree, mobilization costs. The owner's actual daily return must be higher for the higher investment units. Of equal importance is the investment's effect on insurance and tax costs.

Investment costs are determined by design, location of construction or conversion, and when the rig was built or converted. With these variables, it is difficult to look at the original cost of current rigs and to forecast future costs and rentals. In general, barge units currently operating have cost less than the other two types of units. Most of the barges represent initial investments of $3 to $6 million. Current ship units have cost slightly more—in the range of $4 to $7 million. The semisubmersible units have been the most expensive, running from $5 to $20 million.

Drilling and roustabout crew requirements differ little from rig to rig, regardless of the type of unit. Marine crews, on the other hand, are influenced by the type of unit; the self-propelled or ship units are required by law to have larger, more experienced, hence more costly crews.

Daily rental rates (composed of daily operating expenses, return on investment, insurance and taxes) will be in theory the least for the barge units. The ship units with their slightly higher investments and additional crew requirements cost more than the barge units. The semisubmersibles daily cost is highest due to the higher investments which are normally sufficient to offset the larger crew requirements for the ship units.

This comparison of rental rates is strictly a generalization. The range of initial investments overlap and some semisubmersibles built several years ago cost less than some of the barges built more recently.

In selecting a rig for a given job, particularly for a short term, mobilization costs frequently are the determining factor. The actual proximity of the available rig to location of the work is the single most important item. The restricting width of the Panama Canal precludes towing most semisubmersibles by that route and can add major distances to the tow of those vessels.

Ship units have a distinct advantage in mobilization costs because: (1) they travel faster and, hence, spend less time enroute; and (2) they need no high-cost ocean-going tugs to accompany them. Ship units commonly travel at 8 to 12 knots as compared to 7 to 10 knots for the barge units and 4 to 6 knots for the semisubmersibles.

On longer-term contracts, performance begins to assume more importance than mobilization costs. Performance is more affected by layout, equipment, and crew efficiency than all other factors. In comparing the three types of units, it is safe to assume that efficient layout, equipment, and crews can be provided with any type vessel. The comparative aspects really are influenced only by vessel motions. Ship and barge units have natural roll periods in the order of 8 to 14 seconds, depending on the size of the unit. Semisubmersibles have much longer natural roll periods, generally in the range of 18 to 33 seconds. In most open water, the most usual wave periods are in the range of 6 to 10 seconds. This means that semisubmersibles are more frequently out of the range of motion-creating wave periods than are the ship and barge units. The larger ship and barge units are more frequently out of this range than the smaller units.

Semisubmersibles normally have stricter deck-load limits than the ship and barge units. Frequently, semisubmersibles are limited to only one-half the safe load of the others. Ship and barge units also have a higher degree of stability, often in the range of three to one.

In summary, vessel selection is most commonly determined by rig availability and mobilization costs. Oceanographic conditions can, however, influence the performance of the unit and offset higher mobilization costs for the longer-term contracts. Table 6 summarizes comparative data on the three types of floating drill vessels.

Vessel layout

Naval architects, marine engineers, and drilling people have all labored long to improve vessel layout for better operating efficiency and safety. Actual arrangements are as varied as the number of vessels operating. However, certain basic component arrangements seem to have become common to most rigs. The center drill well is located with the center of gravity of the vessel to minimize the effect of rotational motion. The derrick or mast is located over the drill well with the pipe racks adjacent either fore or aft. The rig pumps, the generator, and all noise-generating equipment are usually located in the same section of the vessel as the pipe racks. Crew quarters, galley, recreation areas, and office space are then located in the opposite direction from the center drill well in order to minimize disturbance caused by the operation of equipment.

Most vessels are equipped with two separate administrative

centers, one considered the vessel or marine center and the other the drilling control center. The marine offices are used by the ship's captain or barge master and the marine crew. Usually housed here are the controls and equipment relating to the vessel, i.e., navigation, mooring, trim, and vessel instrumentation. The area serves as the data-collection and record-keeping center on weather, oceanography, deck loads, ballast, and personnel onboard or enroute to or from the vessel.

The drilling control center provides offices for the drilling superintendent, tool pushers, and drilling engineers. Communications are usually controlled from this office since the vast majority of communications have to do with the drilling operation rather than marine operations. Drilling records are maintained in this office. The vessel's position-monitoring equipment is frequently located in the drilling control center also. Most often, office accommodations for the geologists are separate from those of the drilling supervisors and usually placed convenient to mud-logging and core-processing facilities.

Quarters are generally arranged with four men to the stateroom and most vessels have accommodations for 50 to 75 men. Centrally located shower and toilet facilities are also provided. As mentioned previously, these accommodations are remote from the noise areas and are well insulated and air-conditioned. Recreation rooms for off-duty personnel are located in the same area; they usually contain a television, movie equipment, games and reading material. Mess facilities are arranged with cafeteria-type service from a fully equipped galley. Room-size refrigerators and freezer compartments offer perishable-food storage adequate even for the remotest location.

The helicopter landing deck is usually located on top of the quarters and administrative offices. This keeps the helicopters away from the pipe rack area where cranes and other material-handling activities would present flying hazards.

Storage and cargo-handling equipment

As mentioned previously, semisubmersible units are more limited for storage capacity than most ship and barge units. Table 6 compares typical storage capacity for the vessel in drilling position for one of the larger ship units with one of the larger semisubmersible units. Actually, some of the barge units can handle more than 8,000 long tons of materials and supplies. Even the smaller ship or barge units have more capacity than the largest semisubmersibles.

In the ship or barge units, storage is provided near the water line, often even below the water line. Hence, storage weight can easily be countered by ballast.

In the case of the semisubmersible, however, storage is necessarily at considerable height above the water line. While ballast can offset the added weight as far as bouyancy is concerned, it alone cannot be used to control stability.

Most vessels are equipped with at least two deck cranes for onloading, offloading, and handling equipment and material on board. At least one of the cranes is usually in the 45- to 50-ton class and the other somewhat smaller. Both cranes are usually positioned to handle loads over the side and each is within the range of the other. One crane is always positioned near the derrick and can handle loads for the moonpool area. The other generally is located at the far end of the pipe-rack area. Overhead cranes are also provided in the moonpool area for handling the blowout-preventer stack and other subsea gear.

Vessel support facilities

Electrical systems (generators and distribution systems), fuel systems, water systems, ballast systems, and sanitation systems are common to all types of vessels. Water-distillation units are provided on some vessels to supplement water-storage capacity. Mooring winches and conventional mooring-control systems are used on most vessels, but a few have either added to, or replaced, the conventional mooring equipment with dynamic positioning equipment. The mooring system is discussed in detail in a later chapter.

Emergency facilities

Most emergency facilities fall into one of four general categories:
1. Lifesaving
2. Fire protection
3. Pollution control
4. Well control

Lifesaving equipment includes lifeboats, liferafts, lifefloats, life preservers, ring life buoys, distress signals, emergency lighting and emergency communication equipment. The exact facilities provided, their location, arrangement, maintenance, and inspection are carefully controlled by regulatory bodies. Fire-protection equipment likewise is controlled primarily because of its close association to the safety of personnel. Fire-protection equipment includes firemain systems, hoses, pumps, portable and semiportable extinguishers, axes, fire suits, and detection systems.

Pollution-control equipment and practices are less stringently regulated at this time but are extremely important to the industry in general and promise to become more rigidly controlled by governmental agencies. This general category of equipment includes: (1) drip control devices, splash pans, and drains for the entrapment and collection of oil leakage and spills in the engine room or derrick-floor area; (2) washing or cleaning devices for the removal of oil from drilled cuttings; (3) garbage and trash containers for the collection and transportation of solid wastes; (4) chemical dispersants and dispersant systems for the neutralization of contaminants; (5) container booms for preventing the spread of any spillage; and (6) skimmers and pumps for the collection of spilled materials. Pollution control and pollution-control equipment are covered in detail in a subsequent chapter. Well-control equipment and practices are also discussed later.

Vessel inspection and maintenance

Governmental regulatory agencies have set specific inspection and maintenance standards for specific sizes of vessels engaged in specific activities. In the United States, these regulations are under the jurisdiction of the Coast Guard and for most vessels are covered in CG-257. "Rules and Regulations for Cargo and Miscellaneous Vessels". The Coast Guard has recognized the uniqueness of drilling vessels and has made exceptions and special rules for governing certain types of drill vessels.[2]

Drill vessels travel all over the world. They do not return to port at regular intervals and, therefore, are not readily subjected to inspection. In recognizing this, the U. S. Coast Guard has prepared a self-inspection check-off list for mobile drilling units operating overseas. The check-off list is designed to be all-inclusive for all types of mobile drilling units. Hence, some parts of the list do not apply to all units. Each user eliminates from the form those items not applicable.

A preliminary copy of the Coast Guard suggested check-off list is included in this text as Appendix B. It can be used in conjunction with the Preliminary Planning and Rig Evaluation Checklist (Appendix A). The two give a fairly comprehensive evaluation of any drill vessel.

TABLE 3
Ship Units

Vessel Name	Owner
Astragale	Narval
Caldrill I	Fluor Drilling Services
Cyclone	Storm Drilling Co.
Discoverer I	The Offshore Company
Discoverer II	The Offshore Company
Discoverer III	The Offshore Company
Eureka	Sedco, Inc.
E. W. Thornton	Reading and Bates
Exploit	Fluor Drilling Services
Glomar II	Global Marine, Inc.
Glomar III	Global Marine, Inc.
Glomar IV	Global Marine, Inc.
Glomar V	Global Marine, Inc.
Glomar Challenger	Global Marine, Inc.
Glomar Conception	Global Marine, Inc.
Glomar Grand Isle	Global Marine, Inc.
Glomar North Sea	Global Marine, Inc.
Glomar Sirte	Global Marine, Inc.
Glomar Tasman	Global Marine, Inc.
Goldrill IV	Golden Lane Drilling Co.
Goldrill V	Golden Lane Drilling Co.
La Ciencia	Global Marine, Inc.
Navigator	Zapata Offshore Co.
SEDCO 445	Sedco, Inc.

TABLE 4
Barge Units

Vessel Name	Owner
Big John	Atwood Oceanics
C-201	Creole Petroleum
C-202	Creole Petroleum
C-203	Creole Petroleum
C-225	Creole Petroleum
C-226	Creole Petroleum
CUSS I	Global Marine, Inc.
GP-9	Shell de Venezuela
GP-10	Shell de Venezuela
Independencia	Petroleos Mexicanos
Investigator	Zapata Offshore Co.
Nola 3	Zapata Offshore Co.
Nordrill	George Mitchell
Reforma	Perforadora Mexico, SA
Revolucion	Perforadora Mexico, SA
Shiloh	Atwood Oceanics
Western Explorer	Global Marine, Inc.
WODECO I	Fluor Drilling Services
WODECO II	Fluor Drilling Services
WODECO III	Fluor Drilling Services
WODECO IV	Fluor Drilling Services
WODECO V	Fluor Drilling Services
WODECO VI	Fluor Drilling Services
WODECO VII	Fluor Drilling Services
Sonda I	Falcon Seabord Drilling Co.
Tankaigo No. 1	Taiheiya Tonkai Kogyo Co., Ltd.
Terebel	Inst. Francais du Petrole
M/V Torry	No. Sumatra Oil Development Co.
Typhoon	Storm Drilling Co.

TABLE 5
Semisubmersible Units

Vessel Name	Owner
Bluewater No. 2	Santa Fe Marine, Inc.
Bluewater No. 3	Santa Fe Marine, Inc.
Louisiana	Zapata Offshore Co.
Mariner I	Santa Fe Marine, Inc.
Ocean Digger	ODECO
Ocean Driller	ODECO
Ocean Explorer	ODECO
Ocean Prospector	ODECO
Ocean Queen	ODECO
Ocean Traveler	ODECO
Ocean Viking	ODECO
Pentagone 81	Neptune
Scarabeo II	Saipem SpA
Sea Quest	British Petroleum
SEDCO 135	Sedco, Inc.
SEDCO 135A	Sedco, Inc.
SEDCO 135D	Sedco, Inc.
SEDCO 135E	Sedco, Inc.
SEDCO 135F	Sedco, Inc.
SEDCO 135G	Sedco, Inc.
SEDCO H	Sedco, Inc.
SEDCO I	Sedco, Inc.
Sedneth I	Sea Drilling Netherlands, Inc.
Staflo I	Royal Dutch/Shell
Transworld Rig 58	Transworld Drilling Co., Ltd.
Transworld Rig 61	Transworld Drilling Co., Ltd.
White Dragon II	Japan Drilling Co.

TABLE 6

Comparison Of Ship, Barge And Semisubmersible Units

	Ship Units	Barge Units	Semisubmersible
Number operating	29	24	27
Initial cost	$4–7 million	$3–6 million	$5–20 million
Displacement, long tons			
Smallest	450	900	8,000
Largest	36,000	10,600	17,600
Normal	5,000–11,000	3,500–7,500	14,000–17,000
Travel/towing Speed, knots	8-12	7-10	4-6
Natural roll period, sec	8-14	8-14	18-33

Typical storage capacities of larger units, long tons		
Tubular goods	1,270 LT	480 LT
Sack mud and cement	540 (12,000 sk)	230 (5,000 sk)
Bulk mud and cement	230 (5,000 sk)	280 (6,100 sk)
Liquid mud and cement	680 (3,000 bbl)	200 (900 bbl)
Fuel	1,100 (8,150 bbl)	610 (4,460 bbl)
Drilling water	2,400 (15,000 bbl)	1,450 (9,100 bbl)
Potable water	80 (500 bbl)	50 (330 bbl)
Approximate total capacity	6,300	3,300

REFERENCES

1. Marucci, T. F., and McDaniel, D. E.: "Safety of Mobile Offshore Drilling Units". Second Offshore Technology Conferences OTC-1321, April 1970.

2. U. S. Coast Guard Rules and Regulations for Cargo and Miscellaneous Vessels CG 257.

3. Putz, R. P.: "The Motions of Ship-Shaped Vessels in Surface-Wave Environments". 1966 OECON Offshore Conference, Long Beach, Cal.

5

The Drilling System

The drilling system is that part of the industrial equipment on board normally found in land drilling operations. It includes: hoisting equipment; rotating equipment; the drilling-fluid-circulating system and equipment; tubular goods; and the normal rig-floor tools, such as, tongs, slips, and small hand tools. It also includes all the downhole bits, reamers, stablizers, bumper subs, directional tools, fishing tools, and other specialized downhole equipment.

Equipment unique to floating drilling.

Even though most of the drilling-system equipment found on board the vessel is common to its counterpart on land, some of the applications of this equipment are unique. For instance, horizontal motion of the traveling block is restricted by guide rails installed inside the derrick. These hold the traveling block rigidly in the horizontal plane. Mechanically, this is usually accomplished by installing two I-beams inside of the mast. The beams are spaced to provide tracks for rollers mounted on opposite sides of the traveling block. Restraint of the traveling block in turn restricts horizontal movement of the hook and swivel as well.

As stated earlier, the pivotal feature of drilling from a floating vessel is continual vertical movement of the rig floor. Compensation for this motion during drilling is made by installing slip joints, or "bumper subs", in the drill string just above the drill collars. Ideally, this compensates for all of the motion just above the collars and maintains the weight of the collars on the bit.

Experience has shown that it is not actually as effective in compensating for motion as is desired. The subs inherently have some friction and, particularly with high pump pressures, vary the load on the bit. Even with extreme attention by the driller, a change in drilling rate can permit the slack in the bumper subs to "drill-off" or "close-up" depending on whether the rate increases or decreases. In case of a drill-off, upward movement of the vessel will lift the bit off bottom and then impact it on the returning down cycle of the vessel. Ample evidence (broken teeth on bits and surface weight indications) shows that this does occur frequently. Whenever the bumper subs are allowed to "close-up", drill-pipe weight is cyclically applied to the bit and usually as a shock load if the bumper subs open partially on the upward stroke.

Still, up to now, bumper subs are the only method proven workable for motion compensation and do provide a means of operating from a floating vessel. They are expensive to use and maintain, however. One operator has reported that with 251 subs used, average life was 65.2 hours between servicing.[1] In addition to the cost of the subs and their repair, premature failure of a sub can cause the premature pulling of the bit. In case of a structural failure, it can cause a fishing job.

Several manufacturers are currently working on designs and prototypes for surface-operated motion compensators.

One is offering a hydraulic cylinder arrangement for installation between the traveling block and hook. The cylinders
stroke with the heave of the vessel and maintain a constant
hook load. This appears one of the more practical approaches
at this time. One other manufacturer is working on a design
to provide a traction winch that controls the drilling line
and the position of the traveling block in respect to vessel
motion. A third manufacturer is using a piston arrangement
mounted on the deck and attached to the dead line. This
latter system is designed only for emergency use and is
intended to be put into service only when the blowout-
preventer rams are closed and compensation is required for
the drill pipe above the rams.

Motion compensation for wireline work, such as logging
and perforating, has been effectively handled by surface
sheave arrangements. In these, one sheave is attached to a
fixed reference point, such as the top of the marine riser.
Pneumatic pistons suspended from the elevators have also
been used for motion compensation to handle the light loads
of wireline work but do not seem to have any particular
advantage over the sheave arrangement.

In an earlier chapter it was mentioned that Humble Oil
& Refining Co. used a modified torque converter in one of
the earlier core boats, the Humble SM-1. The torque converter was installed between the power unit and the drawworks and permitted the drilling line to feed off or reel in
with the vessel motion. From all reports the system was
effective, but it was used only on shallow core holes. Information is not available on the drilling-line service with this unit.

Many of the earliest floating drilling vessels mounted the
rotary table on a device that permitted the table to gimbal
with the roll and pitch of the vessel. It was felt at that time

the gimbal arrangement was essential if vessel motion became severe. It was intended in such a case to disconnect the kelly and set the drill pipe on the slips, so it would not be unduly stressed at the point of flexure. The actual value of the arrangement is doubtful and no vessels are now constructed with the gimbal. Certainly, in the deeper water, its utility is even more questionable.

In discussing the unique applications of equipment, we return again to the selection of drill pipe for the floating drilling vessel. With motion compensation currently provided just above the drill collars, the entire drill string moves with the motion of the vessel, hence, continually moves up and down in respect to the marine riser, the blowout preventers, the wellhead, and the well casing. Hard-banding should never be used on the drill pipe because of the excessive wear it can cause on the other components.

Well design, drilling practices unique to floating drilling.

Just as in normal land drilling operations, the first step in the design of any new well is a review of available geological information and of previous drilling experience in the area. From these data, estimates of formation pressures and the resulting requirements for drilling-fluid densities are developed. Then, selecting casing-setting depths brings the first influence of the water operation. There is a difference in hydrostatic head at the ocean floor between, on the one hand, the drilling fluid within the marine riser and the sea water outside of the riser on the other. This differential definitely affects the selection of casing-setting depths in order to limit formation fracturing. The effect is small in shallow water. However, protecting against fracturing can become a major factor in the design of wells in water depths greater than

1,000 ft. For any given formation-fracture gradient, the effect of water depth can be readily calculated.

In selecting drilling-fluid densities for the offshore well program, special consideration must also be given to water depth. Suppose the operator customarily provides 300 psi overburden pressure for land operations. Then, additional overburden pressure must be carried for the offshore wells to provide the same degree of protection. The mud density must be selected to provide this 300 psi protection on the basis of the marine riser being disconnected or having developed a leak near the ocean floor. This precaution becomes significant in water depths greater than 600 ft, or with abnormally high mud weights. Say, for example, the riser were disconnected or developed a leak at the bottom while drilling in 600 ft of water with 15.0 lb/gal mud. Overbalance on the formation would be reduced to only 100 psi if a design overburden of only 300 psi were used with the marine riser full of mud. In a water depth of 1,000 ft, under the same design conditions and mud weight, the well would flow if the riser were disconnected.

Special consideration needs also to be given to the drilling-fluid system on the floating vessel. Pit-level devices need to be installed on the center line of the tanks in order to minimize the effect of vessel pitch and roll. It is probable that more accurate gauging could be done with the use of multiple sensors and the integration of the data. Flow devices in the return system are useful only in periods of extremely calm seas, since vessel motion and the corresponding action of the telescopic joint create surges in the rate of flow.

It is a good practice to use a separate small mud tank (about 100 bbl capacity) as a trip tank. It can measure the volumes of mud displaced by the drill pipe on running in the hole and measure the fill-up on trips out. This has proven to be the most accurate method for the early detection of well flows during trips. In fact, even during drilling operations, it is highly desirable to operate with as small a surface volume of active mud as possible. If the surface volumes are limited to 200 to 400 bbl., the odds are increased for faster detection of flow. Moreover, the cost of treating the smaller volume of mud is usually lower.

The driller on board the floater must be even more alert for potential well flows than his counterpart in the land operation. Well flows are not as easy to detect because of surges that occur in the returns due to vessel motion. In addition to being more alert, he has extra steps to consider in the action that he will take if the well does flow. When he adds each single of drill pipe to the drill string, he needs to consider in advance the amount of tide and vessel heave. Only then can he know exactly where his tool joints are in respect to the blowout-preventer stack on the ocean floor. He must be prepared in advance to know exactly where to position the top tool joint in respect to the rotary. This permits him to be certain that, when he closes the rams, he will not close them on a tool joint or just above one. An improperly positioned tool joint can cause damage to the rams. Or, in the case of an upward vessel motion, a tool joint positioned just below the rams could part the drill string.

As a follow-up step to closing the preventers for the control of a well flow, steps should be taken as soon as practical to hang one of the tool joints on the rams. Doing so prevents vessel motion from wearing the sealing element. Advance plans should be made for the support of the drill pipe above the rams once it has been hung off, if needed. In shallow water, up to about 250 ft, it is satisfactory to leave the drill pipe free standing above the preventers. In deeper water, however, air tuggers should be connected at the surface to maintain tension on the drill string.

Float valves should always be used in the bottom of the drill string when drilling through preventers. This permits disconnect at the surface even in an unbalanced mud-pressure condition. Safety valves or stop cocks maintained on the rig floor are a good "extra" for insuring full well control at all times.

Newly hired or proposed rigs

Appendix A, "Preliminary Planning and Rig Evaluation Checklist," was presented in an earlier chapter. It assists in planning and evaluating the equipment for a floating drilling operation. The drilling-equipment section of that appendix should help in evaluating the drilling system.

REFERENCES

1. Harris, L. M.: "Design for Reliability in Floating Drilling Operations". Second Offshore Technology Conference, OTC-1157, April 1970.

6

The Mooring System

The same basic design criterion of providing for safe, efficient operations applies to the mooring system as to other components of the total floating drilling system. The reliability of the mooring system becomes even more important when we consider that all other systems are designed on the premise that the mooring system will retain the vessel on the location within an acceptable range of translational motion.

Most floating-drilling mooring systems in use today are designed to restrict horizontal movement to 10% of water depth. This seems a satisfactory limit for design of the well-control and communications system. Still, it does not require unreasonable components in the mooring system. A mooring-system design to develop high restoring forces early in the displacement and to restrict vessel displacement to only 3% for most environmental forces further increases efficiency in drilling. In other words, the system should be designed to hold the vessel within 10% of water depth for the most severe weather anticipated. Yet, it should restrict the vessel to only 3% for the environmental conditions experienced 95% of the time.

Spread mooring systems represent a large part of the initial vessel investment. The mooring-equipment investment—

including windlasses, chain, wire, anchors and other components—totals about $1 million to equip a vessel to operate in water depths of 1,000 to 1,500 ft, even in a moderate environment. Investments of this size warrant the engineering time needed to develop optimum designs.

Environmental forces

Three environmental forces acting on the vessel are considered in mooring design—the wind, the waves, and the currents. Current forces also act on the marine-riser system and on the mooring lines themselves. Then, too, both wind and waves apply forces to the marine riser and the telescoping joint on a semisubmersible rig.

Extensive calculations are required in determining the magnitude of the forces, in distributing the forces to the individual mooring legs, and in evaluating the behavior of the overall mooring system. The calculations are tedious but can be handled adequately with current technology. The adequacy of the system to perform its function, however, depends not only on the mathematics and these calculations but also on the criteria established for the design.

The assembly and the review of historical environmental data provide the basis for judgement decisions on design winds, waves, currents, and predominant weather directions. These design conditions, properly applied to the vessel configuration, develop the forces required for the mooring-system design.

Wind forces

Wind-tunnel data obtained from tests on a model of the drill vessel are frequently used for determining wind forces, particularly on semisubmersibles and other designs with complex surface areas.

The American Bureau of Shipping has published a method of calculating these data in lieu of wind-tunnel tests.[1] The wind force is determined by the following formula.

$$F = 0.00338 \ A \ V_k{}^2 \ C_h \ C_s$$

Where:

F = the wind force, lb.

A = the projected area, sq ft, of all exposed surfaces in either the upright or heeled condition.

V_k = the wind velocity, knots

C_h = the height coefficient (Table 7)

C_s = the shape coefficient (Table 8)

In calculating wind forces, the following procedure is recommended:

1. On units with columns, the projected area of all columns are to be included; i.e., no shielding allowance should be taken.

2. Areas exposed due to heel, such as underdecks, etc., should be included using the appropriate shape factors.

3. The block projected area of a clustering of deck houses may be used in lieu of calculating each individual area. The shape factor should be assumed to be 1.1.

4. Isolated houses, structural shapes, cranes, etc., should be calculated individually using the appropriate shape factor from Table 8.

5. Open truss work commonly used for derrick towers, booms and certain types of masts may be approximated by taking 30% of the projected block areas of both the front and back sides; i.e., 60% of the block area of one side. The shape factor should be taken in accordance with Table 8.

6. The vertical height should be subdivided approximately in accordance with the values listed in Table 7.

The American Bureau of Shipping (ABS) does not specifically offer guidelines in the selection of wind velocities (V_k) for mooring system design. This value should, however, consider both the average wind velocity and the maximum gust velocity.

TABLE 7
Height Coefficients For Wind Force Calculations
(from ABS)

Height (ft)	C_h
0-50	1.0
50-100	1.1
100-150	1.2
150-200	1.3
200-250	1.4

Note: The height is the vertical distance from the design
   water surface to the center of area of A, as defined.

TABLE 8
Shape Coefficients For Wind Force Calculations
(from ABS)

Shape	C_s
Cylindrical shapes	0.5
Hull (surface type)	1.0
Deck house	1.0
Isolated structural shapes (cranes angles, channels, beams, etc.)	1.5
Under-deck areas (smooth surfaces)	1.0
Under-deck areas (exposed beams and girders)	1.3
Rig derrick (each face)	1.25

The relationship of $V_k = 0.6 \ V_a + 0.4 \ V_g$ (where V_a is average wind velocity and V_g is the maximum gust velocity) gives a reliable value for V_k in mooring calculations.

Wave forces.

Test-model data are frequently used to establish wave forces, just as wind-tunnel data are used for the wind forces. This type of information is particularly helpful to establish loads on semisubmersibles and other irregular shapes.

ABS has outlined theory for the calculation of wave forces for both shallow-water and deep-water operations. A depth of 300 ft is used to differentiate between the two theories.[1] Details of these calculations are included in Appendix C.

The shallow-water method presented is a simplification based on an interpolation between the solitary and Airy theories, and several others. The analysis is based on vertical cylindrical structures and thus may be used for units having structural and stability columns. It can also be used, without serious error, for truss-type legs with non-cylindrical components.

The method also assumes that the structure extends to the bottom of the sea. Where the legs or columns stop short of the bottom, as with a floater, it may be assumed that the forces have diminished greatly at such point, and the non-existent portion below ignored. Or, an adjustment may be made, finding the effective wave height at that distance below the water. Then, another calculation can be made of the imaginary portion below the actual structure, and that subtracted from the original value.

Wave theory for the deep water is a development of the sine-wave theory. It is given by the ABS for use in determining the drag and inertial forces on the underwater portions of drilling units which may operate in water depths greater than 300 ft.

For ship-shaped vessels, wave forces can be estimated more simply based on the relationship of wave period and the ship dimensions.

Where:
   F = the wave force, lb.
   T = the wave period, sec.
   H = the significant wave height, ft.
   L = the length of the vessel, ft.
   B = the beam of the vessel, ft.
   D = the draft of the vessel, ft.
bow forces can be estimated by
   $$F_{bow} = (0.273H^2B^2L) \div T^4$$
where $T \geq 0.332\sqrt{L}$
   $$F_{bow} = (0.273H^2B^2L) \div (0.664\sqrt{L} - T)^4$$
Where $T < 0.332\sqrt{L}$
Beam forces can be estimated by:
   $$F_{beam} = (2.10H^2B^2L) \div T^4$$
Where $T \geq 0.642(B + 2D)^{\frac{1}{2}}$
Or:
   $$F_{beam} = (2.10H^2B^2L) \div [1.28(B + 2D)^{\frac{1}{2}} - T]^4$$
Where:
   $$T < 0.642(B + 2D)^{\frac{1}{2}}$$

Current forces.

As mentioned previously, current puts a load on the mooring system in three ways: the force on the vessel; the force on the riser; and the force on the mooring system itself. Then the total mooring force due to current can be expressed as:

   $$F_t = F_v + F_r + F_m$$

Where:

F_t = total force due to current, lb.

F_v = force on the vessel due to current, lb.

F_r = force on the riser due to current, lb.

F_m = force on mooring components due to current, lb.

In most areas, currents rapidly dissipate below the surface. They are rarely significant below the top few hundred feet of water depth. Because of the relatively small surface area exposed in this short interval, the current forces acting on the riser and the mooring system can often be disregarded. With the relatively large submerged surface area of the drill vessel entirely within this top zone, however, the current loads on the vessel can be quite important.

For ship-shaped vessels, current forces on the vessel can be estimated by[9]:

$$F_{v(beam)} = 0.30 \text{ A } V_c^2$$

and

$$F_{v(bow)} = 0.016 \text{ A } V_c^2$$

where

$F_{v(beam)}$ = force of the current on the beam, lb.

$F_{v(bow)}$ = force of the current on the bow, lb.

A = the wetted surface area, sq ft.

V_c = the current velocity, knots

For semisubmersibles, current forces on the bow or beam of the vessel can be estimated by:

$$F_v = (2.4A_c + 5.7A_f)V_c^2$$

Where:

F_v = current force on the vessel, lb.

A_c = total projected area of all cylindrical members below the surface, sq ft.

A_f = total projected area of all other members below the surface, sq ft.

V_c = current, knots.

When considered, the current forces on the mooring lines may be estimated by:

$$F_m = 1.4 \ A \ V_c^2$$

Where:

F_m = total force on all the mooring lines, lb.

A = projected area of the mooring lines, sq ft. A for wire rope would be the quotient for the number of lines, the suspended length of the lines in feet, and the wire rope diameter in feet. For chain, multiply the chain wire diameter by 3 to allow for the greater exposed area.

V_c = average current, knots.

Current forces on the riser may be estimated by:

$$F_r = 0.01 \ A \ V_c^2$$

Where:

F_r = total force on the riser, lb.

A = exposed area of the riser, sq ft.

V_c = average current, knots.

Mooring patterns.

It is assumed that the vessel will be moored in the most favorable position; i.e., with the predominant and most severe weather to the bow of the vessel. With the vessel heading thus established, the calculated forces can then be used to establish mooring patterns and to calculate individual mooring leg loads.

The selection of the mooring pattern is influenced by the magnitude, direction, and occurrence frequency of the design forces and by the configuration of the vessel. Innumerable patterns have been used for various sets of conditions.[2] Most are symmetrical to some degree.

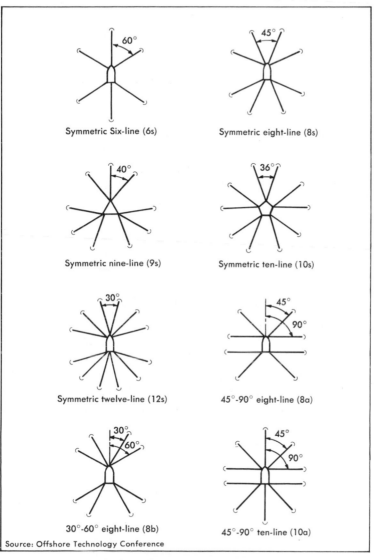

Symmetric Six-line (6s)

Symmetric eight-line (8s)

Symmetric nine-line (9s)

Symmetric ten-line (10s)

Symmetric twelve-line (12s)

45°-90° eight-line (8a)

30°-60° eight-line (8b)

45°-90° ten-line (10a)

Source: Offshore Technology Conference

FIG. 6.1 Spread Mooring Patterns

Several patterns need to be considered and some evaluated as to the optimum arrangement for the vessel and the various conditions of weather. Practical limits of the number of mooring lines to be provided have to be considered, and individual line forces must be calculated for each set of conditions.

The catenary equation.

The catenary is defined as the curve assumed by a perfectly flexible inextensible cord of uniform density and cross section hanging freely from two fixed points, Fig. 6.2 (a). In the spread mooring system, current forces acting on the mooring lines and the stretch in the lines themselves keep the lines from being treated as true catenaries. However, equations based on the catenary curve are sufficiently accurate for floating-drilling mooring designs.

One-half of the catenary curve, Fig. 6.2 (b), gives a good representation of one line of the mooring system. Point A represents the lowest point of the catenary and the point that the line is tangent to the ocean floor. If this line is kept tangent, all of the force on the anchor is in a horizontal direction and the anchor provides maximum holding power. The tension at this point is represented in the free body diagram as Q. Tension F then represents the maximum tension in the line and occurs at the point of fixation on the vessel. The product ws of the length s of the line and the unit weight w of the line in water represents the force of gravity. These three forces—Q, F, and ws—must intersect at a common point to give equilibrium. Terms and units described by Fig. 6.2 (b) are:

F = cable tension at top, lb.
Q = cable tension at point of tangency on bottom, lb.

w = cable weight per foot in water, lb.

s = length of catenary, or free line, ft.

d = vertical distance between free points of the catenary, ft.

k = Q/w, by definition, ft.

y = vertical distance of d+k, ft.

Θ = the angle between the horizontal and force F

x = horizontal distance between free points of catenary, ft.

For the diagram and terminology in Fig. 6.2(b), the equation of the catenary[3] is given by

$$y = (k/2)(e^{x/k}+e^{-x/k}) = k \cosh (x/k)$$

From this equation mathematically we can develop equations particularly useful in mooring design and evalution:

$$k = (F/w)-d$$
$$s = [d(2k+d)]^{\frac{1}{2}}$$
$$x = k \log_e (s+y)(1/k)$$

Thus, environmental forces are established, certain mooring patterns are assumed, and individual line forces are calculated. From these, the equations permit determining the restoring forces, vessel displacements, line sizes, line lengths, and horizontal forces applied to the anchors.

Obviously there are many variables: pattern, line size, line lengths, initial tensions. For this reason, the problem lends itself to computer calculations. This becomes particularly true when deviations are made from the simple single-phase system. Further complications might also be added, such as a two-phase system of chain and wire rope, a sloping ocean bottom, the use of dead weights suspended on the mooring line between the point of tangency and the vessel, or the use of spring buoy systems. The various combinations of these make calculations quite lengthy.

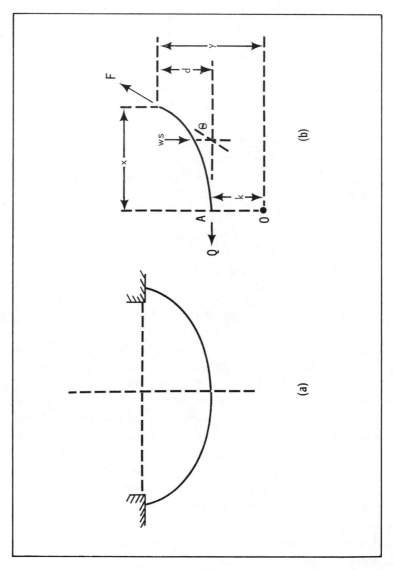

(a)

(b)

FIG. 6.2 The Catenary

64

Design summary

It has now been shown that forces acting on the float-ing-drilling system include: the wind, wave, and current forces acting on the vessel; the current forces acting also on the riser and mooring system; and, for the semisubmersible, wind and wave forces acting on the top of the riser and/or the telescoping joint. The mooring system design then incorpo-rates an arrangement of lines capable of withstanding the maximum forces anticipated and limiting the translational motion of the vessel. Factors affecting the distribution of these forces, the vessel displacement, and the individual loads in each mooring leg will be: number of individual legs; pattern; initial tensions in the system; components of the legs; and water depth.

Complete design first considers the magnitude and direction of the forces loading the system. It then provides: the pattern arrangement; individual component selection; a schedule of line tensions versus anticipated vessel displacements; and recommended initial tensions.

Strengths of the individual lines and all connecting links of the system should be selected with a minimum factor of safety of three based on the strength of new materials. Used materials are satisfactory only if adequately inspected and tested.

Industry experience

Spread mooring systems of the types discussed have been used extensively in floating drilling. The deepest-water expe-rience reported to date is Humble Oil's in the Santa Barbara Channel off the coast of California. The company reports drilling in 1,300 ft of water with some of the anchors set in water depths of 1,500 ft.[4] This was with a combination chain and wire-rope system with 2-in. chain, 2¼-in. wire rope and 20,000-lb anchors.

The U.S. Navy has reported experimental mooring in 15,600 ft. of water with the USNS Josiah Willard Gibbs, T-AGOR-1, a small seaplane tender. This was a single-wire-rope system utilizing a combination of ½-in. and ⅝-in. wire rope and 500-lb Danforth anchor.[5]

Dynamic positioning

The concept of mooring a vessel with propulsion equipment rather than fixed mooring lines is called dynamic positioning. The ability to hold the vessel successfully within acceptable displacements by dynamic positioning depends upon the size, range, and flexibility of the power equipment and on the speed of control and response of the propulsion equipment. Systems of this type are practical only through the use of computers.

Dynamic-positioning equipment has been tested on vessels used for dredging, ocean bottom surveys, and core drilling. Perhaps the best known experience with dynamic positioning has been with the Western Offshore Drilling and Exploration Co.'s Caldrill I and with Global Marine's Challenger.

In general, dynamic positioning is more applicable where short-term position holding (frequent moves) is required, or in water depths beyond the current technology for spread mooring systems. Dynamic positioning offers the advantages of minimum mooring time and free rotation of the vessel for changing weather conditions. The Challenger reported an average mooring time of four hours.[6] Fuel consumption is probably the greatest disadvantage. Also, from a practical standpoint, the magnitude and range of power requirements limits the environmental conditions that can be sustained.

The Offshore Co. has developed and used a turret mooring system. In it, the spread mooring system is connected to a turret around the moonpool and the vessel uses thrusters to change heading. This system seems to combine some of the advantages of both the spread mooring system and the dynamic-positioning system. Its greatest disadvantages are probably in the limits it places on work room around the moonpool and in the installation and maintenance costs on the turret equipment.

Anchors

Three styles of anchors are commonly used in spread mooring systems for floating drilling: the U. S. Navy light-weight type anchor (LWT); the Danforth Anchor; and the stockless anchor. Most frequently, these are 10,000-, 20,000- or 30,000-lb class with proof test loads of 125,000 to 275,000 lb. A newer style anchor—the U. S. Navy stato anchor—is being used to a limited degree in floating drilling. Preliminary reports indicate superior holding power in both sand and mud bottoms.

Ocean-bottom conditions, of course, do affect anchor holding power. It has been found that adjusting fluke angles can improve holding power. Generally, in softer mud and silt bottoms, the larger fluke angles give superior holding power. Optimum setting of the fluke angle for a soft bottom appears to be about 50°. For a sand bottom, a lower fluke angle of 30° to 35° seems best. For unknown bottom, a fluke angle setting somewhere between the two—35° to 40°—should be used.

Some field work has been conducted in predetermining bottom conditions and forecasting anchor holding power prior to arrival of the floating drilling vessel.[7] This is done by measuring the holding power of small (3,000 and 5,000-lb)

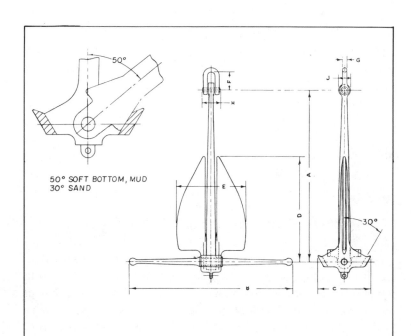

50° SOFT BOTTOM, MUD
30° SAND

ANCHOR WEIGHT	A	B	C	SHACKLE SIZE	PROOF TEST LBS.
3,000	103½	104	66	3	45,000
6,000	124	118	114	4	81,500
10,000	144	137	88	4	125,000
15,000	157	149½	96	4	170,000
20,000	173	164	106	4½	220,000
25,000	186½	206	114	5	249,000

(Source of Illustration: Baldt)

FIG. 6.3 U. S. Navy Lightweight Anchor

68

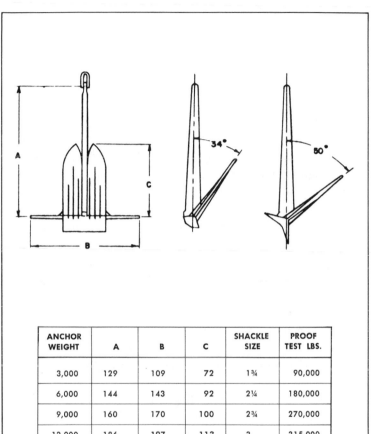

ANCHOR WEIGHT	A	B	C	SHACKLE SIZE	PROOF TEST LBS.
3,000	129	109	72	1¾	90,000
6,000	144	143	92	2¼	180,000
9,000	160	170	100	2¾	270,000
12,000	186	197	113	3	315,000
15,000	204	224	126	3½	360,000
20,000	210	230	129	3¾	435,000
25,000	226	246	137	4	490,000

(Source of Illustration: Baldt)

FIG. 6.4 Anchor Dimensions

69

anchors and extrapolating the data to larger anchors. Results are predictable if bottom conditions are consistent over the area and if the preliminary tests are controlled to provide accurate, applicable data.

Experience has shown that few anchor problems, per se, are encountered with floating drilling. A few simple guidelines will help keep these problems to a minimum:

1. Until additional information is available on the performance of the stato anchor, the U. S. Navy LWT seems to have superior holding power per unit weight for most ocean-bottom conditions.

2. Anchors should be selected with a proof test at least 50% greater than the design holding power required.

3. Anchors should be used as designed, without the addition of mud hooks, added area to flukes, locked flukes, or other modifications that could affect the balance and performance of the anchor.

4. Used anchors with unknown service histories should be carefully inspected for cracks or other indications of impending failure.

5. New anchors should be inspected after four to five years of service. Older anchors should be inspected more frequently, up to every two years for anchors twenty years of age.

6. Anchors should be lowered into position slowly and with the anchor line in slight tension to permit the flukes to dig in as the pendant tension is relaxed.

7. After setting, the anchors should be pre-tested to the maximum design load and reset if not holding.

Anchor piles or drilled-in anchors

In some areas where shale or hard rock bottoms preclude the use of conventional anchors, operators have used either driven or drilled-in anchor piles for mooring. Depending on local conditions and the availability of equipment, these can be set in advance by auxiliary vessels. Or, the drill vessel itself can set them when it arrives on location. Anchor piles when properly designed and installed provide holding power equal to, or greater than, conventional anchors. However, they are usually more expensive to install, particularly if installed by the drill vessel.

Probably because of experience from normal drilling operations, drilling-in anchor piles are usually established by drilling a shallow hole, lowering a casing stub with an anchor line attached, and grout cementing the casing in the hole. The anchor lines established with the casing stubs are short and extend only to the surface of the water. The drill-vessel mooring lines are then attached to these short lines as the vessel is located in drilling position.

Chain

Two types of chain are satisfactory for the loads and service required for floating-drilling operations. These are flash butt-welded stud link chain and Di-Lok chain. Both types are manufactured in sizes from ½-in. up to 4¾-in. wire diameter, but it is seldom that any chain less than 2-in. or greater than 3-in. is used in floating drilling. Di-Lok chain is available in both a regular strength and a super-strength chain. Comparative strengths of the 3-in. chain[8] are:

	Proof test, lb	*Break test, lb*
Flash-welded (Gr. 3)	679,200	970,300
Di-Lok regular	693,000	1,045,000
Di-Lok super	693,000	1,274,900

All chain should be periodically inspected and test loaded to insure reliability. In the design specifications for the mooring system, it was recommended that all components of the system have a factor of safety of three based on the strength of new components. The chain size and type should be selected on this criterion of three times the strength required under the most severe loading anticipated. In-place testing on each location of the maximum anticipated load was also recommended. Shop testing to loads of 150% of the maximum anticipated loads should also be made. The frequency of the shop test depends on the age and service of the chain. In general, new chain should be tested every four to five years initially and as frequently as every two years when it reaches 15 to 20 years of age.

At the time that it is test loaded, it should be visually inspected for wear or other weaknesses. Chain should be downgraded when grip diameters are reduced due to wear by 10%. They should be discarded when reduced by 20%. For example, 2-in. chain with a worn grip diameter of 1.80 in. (10%) reduction could be satisfactory used as a 1.75-in. chain; but when the worn grip diameter reaches 1.60 in. (20% reduction), it should be discarded.

The ABS sets standards for retirement of chain due to elongation. For instance, six links of 2-in. chain measure 52 in. when new. ABS recommends retirement when this six links measures 52.9 in. ABS standards on elongation should be checked for the chain being tested and used as a guideline in its retirement.

All chain should be visually inspected for fatigue cracks, corrosion and other indications of weakness. Defective links should be removed. If there are too many defective links in any one shot of chain, the entire shot should be discarded.

The studs in flash-welded chain should be checked for looseness by striking with a hammer. Loose studs should be welded or the link cut out and discarded. Links with studs missing must be discarded.

The connections on Di-Lok chain should be checked for separation between the male and female sections. If gaps are found that exceed about 1/32-in. thickness and ½-in. depth, the link should be removed. Longitudinal cracks in the female section are an indication of weakness and sufficient basis also for discarding links.

Inspected shots of chain should be identified for future reference. These identifications should *not* be made by stamping a link as this could cause stress concentrations. An identifying reference number, the proof test and date can be stamped on thin metal strips and connected around the end link of chain.

Wire rope for anchor-line service

For smaller drill vessels and even for larger units operating in deep water, wire rope is commonly used in the mooring system. It may be used either as a pure wire-rope system or as a combination wire-rope and chain system. A 6x19 Seale, right regular-lay, IWRC, improved-plow-steel, galvanized wire rope is recommended for this service.

As with anchor chain, wire-rope diameters should be selected to provide three times the strength required under the most severe loading expected. In-place testing on each location, to the maximum anticipated load, should be a regular procedure.

On every rig move, the wire rope should be inspected and lubricated with a rust-inhibiting oil or grease. Lubricant[5] can be applied with a hand brush as the line is reeled into the vessel, or by means of a box-type lubricator. Lubrication serves the dual purpose of protecting the wire rope from corrosion and of lubricating between the wires.

At the same time the line is lubricated, it should be visually inspected for broken wires, reduced diameter, corrosion, and kinks, mashing or other physical damage.

Wire should be replaced whenever the number of broken wires per lay length of the wire equals about 10% of the wires in the line (exclusive of the core). For example, a 6x19 rope has 114 wires (6x19 = 114). It should be retired from service when about 11 strands are broken per lay length of wire.

Reduction in wire diameter is also a measure of the loss of strength of wire ropes. Diameter is reduced by elongation, wear, and corrosion. In general, a line should be replaced if the diameter is reduced by as much as 6%. For example, a 2-in. line reduced by ⅛ in. (0.125 in.) is reduced by 6.25% and should be replaced. A 3-in. line reduced by 5/32 in. (0.15625 in.) is reduced by 5.2% and is nearing time for replacement.

Kinked or cut lines should always be replaced, or cut off if the damage is near one end. Replacement for corrosion damage depends on the amount of cross-section area missing and can usually be evaluated by wire diameter. Mashed wire generally does not cause a reduction in tensile strength and can be satisfactorily used if not in a flexing position.

The end of the wire rope used in mooring service (at the point where it is socketed) is subject to the most severe fatigue stressing. For this reason, the line should be cut off about 15 ft. periodically and resocketed. In severe service, this should be done as frequently as every rig move. In fairly calm water, a frequency of 18 months should be satisfactory. Diameter measurements and counts of broken wires are also good guides for the cut-off and resocketing operation.

Wire rope for pendants

Wire rope used as pendants is subjected to extremely severe service and is frequently damaged during the handling of anchors. Since corrosion is not the usual problem, galvanizing is not justified. Otherwise, the same 6x19 Seale, right regular-lay, IWRC, improved-plow-steel wire rope is recommended for this service as was for the anchor lines. Diameter chosen depends on the water depth and the loads to be handled by the pendant. The same factor of safety of three applies to pendants, as well. The length of line should exceed the water depth by at least 25% to facilitate tensioning of the anchors during the landing process.

The pendants should be inspected and lubricated on every rig move, just as the anchor lines, and should be replaced by the same guidelines.

Connecting links, shackles, sockets

The selection of connecting links, shackles, and sockets should receive the same engineering attention as other components of the mooring system. Historically, they cause a significant number of failures. Care should be exercised that only forged-steel components are used. Maximum strengths should be provided. Each unit should be sand-blasted and magnetically inspected at least once a year. Even new units should have this inspection before being placed in service. The units should be test loaded the same as chain and discarded with wear reductions of 10%. Tests have shown that the life expectancy of connecting links and shackles will frequently be only 30 to 50% of chain. In all cases, connecting links, shackles and sockets should be discarded after ten years of service. Identifying marks for these units may be made on the same type of thin metal strips recommended for the chain, or they may be made on lead seals or studs. Marks should never be stenciled on the body proper because of the possibility of setting up stress concentrations.

Buoys, lights, reflectors

The principal use of buoys in floating-drilling operations is to mark anchors set in spread mooring systems and to facilitate recovery of pendant lines and anchors. Occasionally, buoys are set near the drill vessel as a mooring point for a supply or standby vessel. Size of the buoy depends on the buoyancy required to support the pendant.

Design is not critical and numerous types of buoys are in use. A cylindrical-shaped, welded low-carbon-steel construction provides a sturdy, long-life unit that is easy to handle. The unit should have strong padeyes on top and bottom for attaching handling lines and pendants. The buoys should be filled with a polyurethane foam to maintain buoyancy even though the shell be punctured or otherwise damaged.

Buoys should be painted for easy detection (orange and white) and when used in areas of heavy traffic should have lights or radar reflectors. Lights, when installed, should be checked daily to verify operation and inoperative lights replaced without delay. Photoelectric cells which operate the lights only at night can prolong battery life.

Wire-rope windlasses, chain wildcats

Drill vessels moored with a pure wire-rope system or a combination wire-rope and chain system are generally equipped with wire-rope windlasses. These serve both to tension the mooring legs and to handle and store the wire rope.

In the combination system, the chain is normally handled exclusively by a support vessel.

Size of the wire-rope windlasses depends on: maximum tension requirements; and storage capacity required for the wire rope when retrieved to the vessel. Dual-drum units are perhaps the most common, but both single-drum and quadruple-drum units are used to some extent.

For pure chain systems, the wire-rope windlasses are replaced by chain wildcat windlasses and the chain is handled by the drill vessel. The wildcat differs from the wire-rope windlass in that the drum is formed to fit the chain links. Its cog-type gear engages and tensions the chain. Unlike the wire-rope windlass, the wildcat windlass does not provide storage for the chain; chain is normally stored in lockers convenient to and below the wildcat units.

REFERENCES

1. American Bureau of Shipping: Rules for Building and Classing Offshore Mobile Drilling Units, 1968.

2. Adams, R. B.: "Analysis of Spread Moorings by Dimensionless Functions". First Offshore Technology Conference, OTC 1077, May 1969.

3. Chambers, S. D., and Faires, V. M.: Analytic Mechanics, The Macmillan Company, New York, 1946.

4. Stanton, P. N., Tidwell, D. R. and Lloyd, J. R.: "Sloping Sea Floors". Second Offshore Technology Conference, OTC 1158, April 1970.

5. Beck, H. C. and Ess, J. O.: "Deep Sea Anchoring". U. S. Government, Cataloged by ASTIA, AD No. 290610, July 1962.

6. Schneider, W. P.: "Dynamic Positioning Systems". First Offshore Technology Conference, OTC 1094, May 1969.

7. (a) Watts, J. S., and Faulkner, R. E.: "A Performance Review of the SEDCO 135-F Semisubmersible Drilling Vessel". 1968 Technical Meeting, Petroleum Society of CIM, Calgary. (b) Cole, M. W., Jr., and Beck, R. W.: "Small-Anchor Tests to Predict Full Scale Holding Power". 1969 Annual Fall Meeting of Society of Petroleum Engineers of AIME, SPE-2637.

8. Baldt Anchor & Chain Corporation, Chester, Pennsylvania, Catalog No. 29E.

9. Graham, John R.: "A Discussion of Problems and Knowledge Concerning Station Keeping in the Open Sea". Ocean Industry magazine, August 1966.

7

Auxiliary Vessels

One of the trade journals has estimated that nearly 5,000 auxiliary vessels service the offshore drilling and production activities of the petroleum industry.[1] These units include crewboats, utility vessels, supply vessels, tugs and towboats, seismic-exploration and oceanographic vessels, and cargo barges. Of this total, probably 450 to 500 vessels actively support floating-drilling operations.

Crewboats

With the advent of offshore petroleum operations, the industry turned to existing equipment to support the operation wherever possible. For crewboats this meant the use of large speedboats and yachts. Today, vessels are being built for the specific purpose of moving personnel between the rig and the shore base and to move them as efficiently and safely as possible. Most crewboats are in the 65- to 85-ft. class and are capable of carrying 25 to 50 passengers at speeds of about 25 knots. Most modern vessels also have a small aft deck capable of handling small cargo, up to about five tons.

Supply vessels

Like the crewboat, most early supply vessels were adaptations of locally available craft. Many were war-surplus LCT's

and some of the earlier vessels designed for the specific purpose of supplying the offshore petroleum industry still copied
this type of craft.

In 1954 a new type of supply vessel came into being and
was to set the pattern for offshore supply vessels that were
to come.[2] These vessels had all of the crew quarters forward
and a large deck area aft for the carriage of deck cargo. In
general, these vessels have been characterized by a low
overall-length-to-beam ratio and a high beam-to-depth ratio.
While length overall may vary from 80 to 180 ft, beams are
seldom less than 32 ft or more than 38 ft. Depths, too, are
restricted to a range of 10 to 14 ft.

These vessels can usually handle 300 to 600 tons of deck
cargo and some are installing below-deck bulk storage for
cement and drilling-mud supplies. Power for these vessels
ranges from 650 to 3,000 hp and is capable of propelling the
vessels at cruising speeds up to about 12 knots.

Tugs, towboats

Ocean-going tugs range in size from 2,000 to 11,000 hp,
but the vast majority used by the petroleum industry are
4,000 to 4,500 hp. The combination tug-supply vessel is
becoming increasingly popular for use with floating vessels
operating in remote areas. These tug-supply vessels meet all
of the requirements of the conventional supply vessel, but
their additional horsepower permits them to tow the drill
vessel.

Special-purpose vessels

Several special vessels have been designed and built specifically to serve offshore drilling. Some have been particularly
helpful to floating drilling vessels. Anchor-handling equip-

ment has been added to a number of supply vessels to equip them for setting and retrieving anchors and chain during mooring operations. Some of these vessels are also equipped with wildcat windlass and chain lockers for handling large quantities of chain.

One service company operating in the Gulf of Mexico off Louisiana has outfitted a vessel for cementing, acidizing, sand consolidation and related activities.[3] This vessel is equipped with pumps, blending equipment, storage capacity for materials, and quarters for service and supervisory personnel.

A test barge has recently been outfitted with complete facilities for handling well-test fluids, for measuring produced volumes, and for disposing of the petroleum products by burning. This unit also has quarters for personnel involved in the activity.

Other special-purpose craft (either designed or built) provide: diving service, including facilities for handling diving bells and small submersibles; ocean-floor core sampling; oceanographic-data gathering; and numerous other services related directly or indirectly to floating drilling.

REFERENCES

1. Alderice, R.: "Offshore Work Fleet Gives Mobility to Oil Industry". Offshore magazine, June 5, 1970, pg 44.

2. Mok, U., and Hill, R. C.: "On the Design of Offshore Supply Vessels". Second Offshore Technology Conference, OTC 1161, April 1970.

3. "New Service Boat Launched". Offshore Magazine, November 1970, pg 34.

8

Well-Control
and Communications
System

The well-control and communications system provides for control of, and communication with, the well bore during drilling operations.[1] It includes such components as: well casing; wellhead; subsea blowout-preventer stack; flexible joint; marine riser; kill and choke lines; telescoping joint; tensioning system; and all of the controls, guide lines and tools necessary for the installation and operation of the system components.

Basic design criteria

Individual components of the well-control and communications system have individual specifications and standards for selection. Still, the overall system and single components have certain basic design criteria in common. First of these, common to the entire floating drilling system, is the requirement for full safety and maximum reliability. The components of the well-control and communications system need to be of lightweight, simplified construction and require minimum installation time with minimum risk. They should be designed

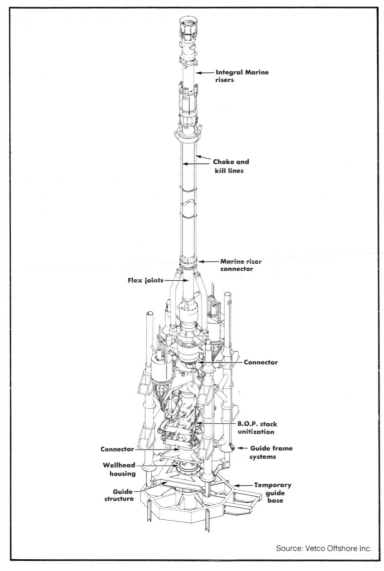

Integral Marine risers

Choke and kill lines

Marine riser connector

Flex joints

Connector

B.O.P. stack unitization

Connector

Guide frame systems

Wellhead housing

Temporary guide base

Guide structure

FIG. 8.1 Well Control and Communications System

for a completely diverless operation and should require minimum maintenance.

Redundancy is the key to satisfying these requirements. Redundancy implies that alternate methods of well control and operating subsea equipment will be available in the event of failure of any one component or group of components. It is provided in the design for this purpose. Care is necessary that the redundancy feature not be misused. That is, if one piece of equipment becomes inoperative or troublesome, drilling operations should be suspended until repairs or replacements have been made. The redundancy feature should not be utilized to continue normal operations in lieu of maintenance.

The subsequent chapters discuss in detail the individual components of the well-control and communication system and provide basic standards for design of the system and for selection of equipment.

REFERENCES

1. Harris, L. M., and Ilfrey, W. T.: "Drilling in 1300 Feet of Water—Santa Barbara Channel, California". First Offshore Technology Conference, OTC-1018, May 1969.

9

Subsea Guide Bases, Guide Structures, Casing and Wellhead Equipment

Most floating-drilling equipment is simply land-drilling equipment modified or adapted to the conditions of the floating operation. Some equipment, like various guidance devices, is highly specialized, however, and has no counterpart in a land operation.

Wellhead equipment is similar in many respects to its land counterpart. It differs principally in that it must be remotely landed and used on the ocean floor. Individual casing strings, too, must be remotely landed, sealed, and released without the benefit of visual inspection and hand operation.

Casing-setting depths

Most casing-design practices used for wells on land apply directly to wells drilled offshore. Casing-settings depths are selected on the basis of: (1) protecting fresh water sands; (2) isolating lost-circulation zones; and (3) restricting the open well bore to formations capable of withstanding the required drilling-fluid densities and the resulting hydrostatic pressures. As to (3), there may also be a need to contain a well flow

85

and the increased pressures imposed on the formation by gas in the well bore.[1]

` All of these considerations are equally important in casing design for floating operations. Water depth affects overburden pressure (hence, fracturing pressure) and requires special consideration for the offshore operations.

The fracturing pressure of the rock surrounding the bore hole is a function of overburden pressure. For land operations, that overburden pressure approximates one psi per foot of depth. In a water operation, this overburden pressure is reduced for the section from the water surface to the ocean floor. This is not significant in selecting casing-setting depths for shallow water or for deeper casing strings. It can become extremely important in water depths greater than 1,000 ft, particularly for those casing strings that penetrate the ocean floor by less than 3,000 ft. Then, too, in operations offshore, the height of the drilling-fluid returns above the surface of the water (or overburden) is generally more than on land operations. This is particularly the case on semisubmersibles where the distance could be as much as 60 ft. This also has an adverse influence on the magnitude of the drilling-fluid hydrostatic pressure in relation to fracturing pressure.

In calculating the effect of water depth, we need first to equate the hydrostatic pressure of the drilling fluid to that of the formation fracture pressure in terms of drilling-fluid density, well depth, and water depth. Hydrostatic pressure of the drilling fluid is:

$P_m = 0.052g \, (d+h+e)$

Where:

P_m = hydrostatic pressure of the drilling fluid, psi.

g   = density of the drilling fluid, lb/gal.

d   = depth of the well below the ocean floor, ft.

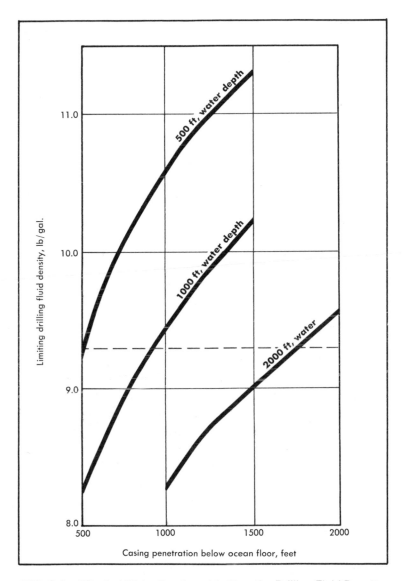

FIG. 9.1   Effect of Water Depth on Limiting the Drilling Fluid Density

87

h   = water depth, ft.

e   = elevation of the drilling-fluid return overflow above
       the water surface, ft (usually about 60 ft for semisub-
       mersibles and 25 ft for ship shapes).

Formation fracture pressure is:

   $P_f = C(wh + rd)$

Where:

   P_f = fracture pressure of the formation, psi.

   C = percent of theoretical overburden pressure where the
        fracture initiates (usually about 70%).

   w = fluid head of sea water in psi/ft (0.445).

   h = water depth, ft.

   r = weight of overburden rock, psi/ft (1.0).

   d = depth of the well below the ocean floor, ft.

   In theory, when $P_f = P_m$ the formation will fracture. Hence,
equation can be set equal to equation:

   $0.052g(d + h + e) = C(wh + rd)$

For any given drill vessel and water depth and with known
and accepted emphirical values, the casing requirements can
be established as a function of mud density. Fig. 9.1 illustrates
the effect of water depth for one set of conditions. The curves
included therein were developed for: a semisubmersible rig
with the drilling-fluid return overflow 60 ft above the water
level; and an assumed fracture gradient of 70% of theoretical
overburden pressure. Actually, the height of the overflow has
only minor effect and the assumed fracture gradient is reasonably good throughout the world. Hence, these curves are
accurate enough for most situations.

Temporary guide base*

Once the drill vessel has been positioned on location, the
temporary guide base is the initial piece of equipment lowered

to the ocean floor. It serves the dual purpose of: (1) providing the anchor for the four guide lines; and (2) ultimately being the foundation for the permanent guide structure.

The temporary guide base is equipped with a J-slot and is run with a J-type tool on drill pipe. The base also has sawtooth legs to penetrate the ocean floor and prevent rotation either during disconnect from the J-slot or subsequent drilling operations. Provision is made for additional cables which can guide a television camera for monitoring subsequent underwater operations.

As the guide base is readied for installation, weight material is added to compartments in the base. Enough is used to give sufficient weight for tensioning the guide lines and for the operations to follow. The guide base itself weighs about 8,000 lb. About 20,000 to 25,000 lb of weight material are added to that. Sacked barite is normally used for the weighting.

The foundation pile

The first casing or conductor string set is generally 30- or 36-in. OD pipe and is commonly referred to as the foundation pile. It is designed to prevent sloughing of the ocean-floor formations and to give structural support for the permanent guide structure and the subsea blowout-preventer stack subsequently installed. About 100 ft of this casing is normally used but this varies depending on ocean-floor conditions.

---

*Different manufacturers of subsea systems have minor differences in the sequence of operations and features on individual components. Even individual manufacturers offer variations tailored to individual customer requirements. The system discussed in this text is that of Vetco Offshore Industries, Inc. and is commonly used world-wide. Where significant features differ between manufacturers, these differences have been noted and discussed.

Where more than one joint is used, the joints can be connected by a quick connect-disconnect, threadless tool joint designed to eliminate time-consuming welding.

After the temporary guide base has been installed, a utility guide frame is used on the guide lines to guide the bit from the drill vessel to the ocean floor. As the bit enters the center opening of the temporary guide base the utility guide frame comes to rest on the guide base. It can be left in this position until the bit is retrieved. The hole for the foundation pile is drilled with sea water and the return fluid with the drilled cuttings is spilled onto the ocean floor. On completion of drilling, the hole is normally filled with a gel-water mud to prevent sloughing and fill. The bit is then pulled, and the foundation pile readied for installation.

The shoe of the foundation pile is equipped with a break-away guide frame. These arms ride in the guide lines that will guide the bottom of the pile into the center opening of the temporary guide base. They shear away as the shoe enters the hole. To the top of the foundation pile is attached the foundation-pile housing assembly which supports the foundation pile and provides a landing base for the next casing string. The permanent guide structure is attached around the housing and is run at the same time as the foundation pile.

The foundation pile is cemented in place by displacing cement down the drill-pipe handling string and through a stinger string extending through the foundation pile.

Annulus return fluids are spilled onto the ocean floor and can be monitored with the television camera during cementing. The cement is recognizable and the quantity of cement returns easily estimated.

A variety of running tools are available for connecting the foundation-pile housing to the drill-pipe handling string. Left-hand thread, hydraulic, and drop dart-type releases are all commonly used.

The permanent guide structure

The permanent guide structure is attached to, and installed with, the foundation pile. When the foundation pile is set and cemented, a permanent structural base has been established on the ocean floor for further operations. At this point the unit still has no pressure integrity in that only one short section of pipe has been installed below the ocean floor.

The permanent guide structure carries a bottom gimbal base which seats in the funnel section of the temporary guide base and thereby guarantees vertical positioning of the guide posts. The guide posts are normally about 8 in. in diameter and vary in length from 10 to 20 ft depending on the height and arrangement of the subsea blowout-preventer stack. The posts are of a slotted-tube-type construction; split and hinged guide-line traps at the top and bottom of the posts provide for easy installation on the guide lines. The external configuration of the post heads permits the use of a remotely operated connector to reestablish broken guide lines, if needed. In actual operations, guide lines have broken in water depths greater than 1,000 ft. By cutting them off at the top of the guide posts, it has been possible to reestablish the lines without the aid of divers.

Spacing of guide-line posts has varied according to individual customer requirements. In general, the industry seems to be standardizing on a 6-ft radius.

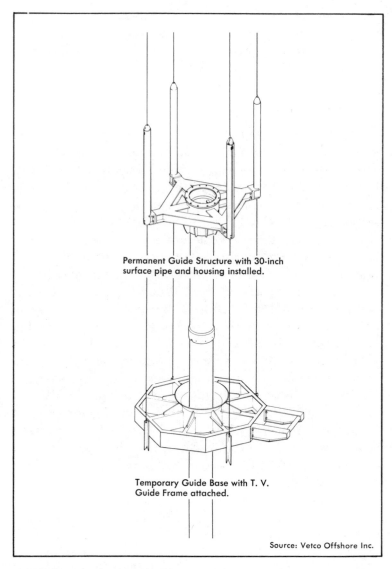

Permanent Guide Structure with 30-inch
surface pipe and housing installed.

Temporary Guide Base with T. V.
Guide Frame attached.

FIG. 9.2   Temporary Guide Base and Permanent Guide Structure

The conductor casing

Although casing programs and individual well requirements vary, a 20-in. conductor casing is commonly set through the 30-in. foundation pile. These pipe sizes seem accepted throughout the industry regardless of the size of the subsequent surface, intermediate, and production casing strings.

The hole for the 20-in. conductor is drilled in a manner similar to that used for the foundation pile. The bit is guided to the top of the base with a utility guide frame, the hole is drilled without blowout preventers, and the returns are taken to the ocean floor.

The hole is filled with a gel-water fluid, and the bit is removed preparatory to running the conductor pipe. The shoe of the conductor is guided with break-away guide arms, just as on the foundation pile.

The wellhead is installed on top of, and run with, the conductor pipe. It is designed with ribbed extensions to fit into the foundation-pile housing. Provision is made inside of the head to install the required number of casing hangers for all future casing and tubing strings. The upper part of the wellhead has an external configuration designed to accept and latch the connector of the subsea blowout-preventer stack. An AX ring gasket provides a metal-to-metal seal between the connector and the wellhead and thus establishes pressure integrity for future operations.

The wellhead and conductor pipe are lowered into position with a hydraulically operated connector installed on the drill string and attached to the wellhead. As the wellhead seats on the previously set permanent guide structure and foundation-pile housing, it is latched and rigidly attached to the housing.

A stinger string extending the drill pipe to the shoe of the conductor is commonly used in displacing cement. Again, cement returns are taken to the ocean floor and can be observed with the television camera.

Casing hangers

Following the running and cementing of the conductor string with the wellhead, the subsea blowout-preventer assembly is installed and used on all deeper drilling. Future casing and tubing strings are run through the preventers and landed in the wellhead. The wellhead then acts as intermediate between the casing strings and the subsea blowout-preventer stack, providing both a means of support and a means of connection. The wellhead can be supplied to accommodate as many additional casing and tubing strings as required.

The casing hangers stack vertically in the wellhead and are normally latched in the wellhead housing. The hangers have a flow-by area that permits fluid circulation during cementing. Afterwards, the area is sealed with a resilient packing. Running tools are available that permit the sealing operations subsequent to the cementing without an additional trip of the tools.

Casing-setting operations are normally designed with the hanger located in the casing string at a predetermined point. This necessitates landing of the casing at the predetermined depth. In an emergency, or if the casing is stuck during running, alternate methods of landing the casing are available. If clearances are sufficient, the casing can be disconnected below the wellhead. The casing hanger can then be reassembled at a new point and rerun with a pack-off overshot to connect to the lower section of the casing. If preferred, the

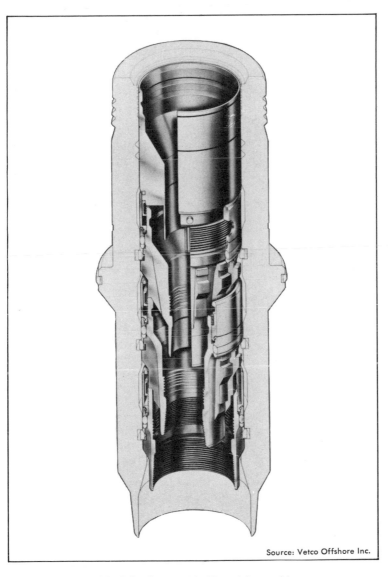

FIG. 9.3 Subsea Wellhead Assembly

95

casing can be cut in the wellhead and a slip-type emergency hanger installed around the casing. This second alternate has been used and is practical, although time-consuming in most cases.

Since the hanger assemblies are designed to latch and seal in the wellhead housing, protection is needed for the sealing surfaces in the wellhead during drilling. Removable sleeve-type bore protectors are installed subsequent to each casing-string installation. These are left in place until the next casing string is to be run.

REFERENCES

1. Goins, W. C.: Blowout Prevention, Gulf Publishing Company, Houston, Texas, 1969.

10

Subsea Blowout-
Preventer Stack

Purpose of the blowout-preventer system in floating drilling is identical with that in conventional land drilling. This is, it provides control when well flows develop and provides a means of circulating, conditioning, and returning the well bore to a static, unpressured condition. Specifically for floating operations, the arrangement should provide reliable means of: (1) closing in around the drill pipe and circulating a conventional well kick; (2) being able to sustain these well conditions over a prolonged period; (3) hanging off the drill pipe, closing in the well and moving the drill vessel off location; (4) being able to reestablish the drill vessel on location, monitor and circulate the well prior to reestablishing the drill pipe; and (5) having alternate methods available with the failure of any one function.

Arrangement of preventers and side outlets

The specific arrangement of preventers and the side outlets has been a subject of much discussion. Most arrangements have advantages and disadvantages. Those recommended here have been selected considering the specific nature of the floating operation. Some of the recommendations definitely do not apply to an operation either on land or on a fixed or mobile platform offshore.

For exploratory drilling, the subsea blowout-preventer stack should include one Hydril-type preventer and four ram-type preventers. It should have two side outlets for the choke and kill-line connections. They should be arranged, from top to bottom:

1. Marine riser connector
2. Hydril-type preventer
3. Ram-type preventer with blind-shear rams
4. Side outlet
5. Ram-type preventers with pipe rams
6. Ram-type preventers with pipe rams
7. Side outlet
8. Ram-type preventers with pipe rams
9. Wellhead connector

All preventers should have a pressure rating in excess of the maximum that could be expected on the wellhead. Both the middle pipe rams (No. 6 above) and the lower pipe rams (No. 8) should be equipped with rams of the same size as the drill pipe most frequently used. The upper pipe rams (No. 5) may be equipped with rams of a smaller size when plans include either drilling the last part of the hole or drill-stem testing with a smaller-size pipe.

All of the pipe rams should be equipped with locking devices to insure that they remain closed even with the loss of hydraulic control pressure. On the Cameron Iron Works Type U preventer, the rams can be locked with a hydraulically-operated, wedge-type mechanism. This blocks movement of the rams until unlatching hydraulic pressure is applied. On the Shaffer Tool Works Type LWS preventer, the rams are mechanically latched into place as the rams are closed. They

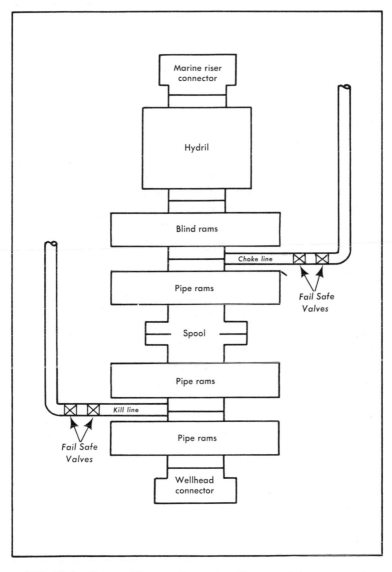

FIG. 10.1 Subsea Blowout Preventer: Equipment Arrangement

are unlatched by applying opening pressure to the ram-open side in the hydraulic system.

The Hydril-type preventer should be equipped with a remotely operated regulator to permit adjustments in the pressure of the unit's hydraulic-power-fluid supply. It should also be equipped for automatic actuation in case of failure of the control system. This latter feature is discussed in more detail in the chapter on the control system.

Clamp connections are preferred over flange connections for the entire subsea blowout-preventer stack. Clamp connections are more compact and help reduce overall height and weight of the subsea unit. They are less likely to loosen and are more easily checked and tightened during the inspection and servicing of the blowout-preventer stack.

Hydraulically operated connectors

In most floating drilling systems, identical connectors are used on the top and on the bottom of the subsea blowout-preventer stack. The upper one connects the marine-riser system to the blowout-preventer stack. The lower one connects the stack to the wellhead. Both connectors should be of the same working pressure as the blowout preventers. The lower connector has to be of this rating since it is always exposed to wellhead pressure. Under normal operations, the upper connector is never exposed to pressure greater than that difference of hydrostatic heads between the drilling mud and the sea water at the ocean floor. The upper connector does need to be of the full rated pressure of the stack, however. This makes it possible to add an additional preventer, or stack, on top of the original stack in an emergency.

Two types of connectors are in common use. Basic difference is in the manner in which they mate with, and attach

FIG. 10.2 Hydraulically-Operated Connectors (Mandrel-type)

101

to, the wellhead (or upper stack connection). In one case, the wellhead is designed as a mandrel with an external configuration providing grooves for the locking dogs of the connector. The connector, attached to the bottom of the blowout-preventer stack, rides over the mandrel, is aligned and is latched by dog segments moved to the lock position by the actuation of a hydraulically positioned cam ring. An AX metal gasket provides a metal-to-metal seal between the two elements.

The second type, known as a collet connector, mates and seals on a wellhead with a hub design. This connector has multiple clamp segments that form a funnel-type configuration to guide and align the connector as it is moved into position. This type connector is also latched by the actuation of a hydraulically positioned ring closing on a locking taper and sealed with an AX metal gasket.

Both types are designed to remain latched even after the loss of hydraulic power. Both types are available with increased hydraulic advantage for unlatching to minimize releasing difficulty, and both are available with mechanical override features, if desired.

In general, both types have performed well. On adequately designed guidance systems, there should be no particular advantage for either in original alignment and latching. Field reports indicate that the mandrel type are usually easier to disconnect during periods of adverse weather and misalignment.

Some manufacturers recommend an AX gasket with a resilient, or rubber, coating for use with their connector. The resilient coating makes it possible frequently to provide a seal that will hold a test pressure even though the ring seat is scarred, corroded, or otherwise damaged. It may seal even with a certain amount of trash or foreign material between the gasket and the seat. The ability of this resilient seal to hold pressure during a short test period can be misleading and can create an unnecessary hazard. Resilient seals should never be used on a routine basis and should be used in an emergency only after weighing all risks involved.

Kill and choke-line valves

Dual hydraulically operated, fail-safe valves are recommended for each side outlet serving as either the kill or choke line. Side outlets are known areas for sand cutting and erosion. For this reason, these valves should be positioned as close to the blowout-preventer stack as possible and with a minimum of connections between the stack and the valves. At least one of the valves on each line should be connected directly to the stack and before the fluid flow path makes a turn. It would be better if both could be located before this turn in the flow path. However, there are width restrictions on the stack and the valves are vulnerable to being broken off if they extend too far. For these reasons, one valve is normally mounted directly on the stack and the second after an ell or turn in the flow path.

In the selection of valves and valve operators for this service, two pecularities of the installation must be considered. First, the valves are to be used in varying water depths. Hence, they must have operators where the effect of the hydrostatic head of the power fluid is counter-balanced by the effect of the hydrostatic head of the sea water, and vice versa. Secondly, when two valves are installed close together in the same line, there is the possibility of liquid locking between the valves. All fail-safe valves, regardless of design, exhaust some fluid from the body cavity (or the balanced-pressure

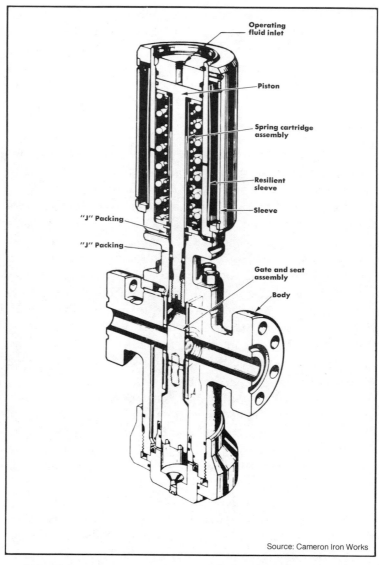

Operating
fluid inlet

Piston

Spring cartridge
assembly

Resilient
sleeve

Sleeve

"J" Packing

"J" Packing

Gate and seat
assembly

Body

FIG. 10.3   Hydraulically-Opened Fail-Safe Choke or Kill Valve

chamber for valves with a balanced stem). Thus the line must be capable of receiving this fluid. Otherwise, the valve will not open. Liquid locking can be avoided by: (1) having upstream valve seal in only one direction; (2) providing a surge chamber; or (3) venting the body cavities or balanced-pressure chamber to the line. When the body cavities are vented, the upstream valve must be vented upstream and the downstream valve vented downstream. The latter method is preferred for floating drilling operations.

OPERATION OF THE SUBSEA BLOWOUT-PREVENTER STACK

At the beginning of this chapter, it was specified that the arrangement of equipment on the subsea blowout-preventer stack must provide for reliable means of performing certain operations for various sets of conditions. Specifically, it should provide for:

1. Closing in around the drill pipe and circulating a conventional kick. This can be accomplished by closing the Hydril-type preventer, landing a tool joint on the middle pipe rams, and circulating. If the alternate drill-pipe size is in use, it would require landing on the upper pipe rams instead. If a hang-off tool is available and if desired, the hang-off tool can be installed and the drill string landed on the wellhead.

2. Being able to sustain these conditions over a prolonged period. With the drill pipe landed and all motion of the drill string through the preventers eliminated, the conditions can be sustained over a prolonged period.

3. Hanging off the drill pipe, closing in the well and moving the drill vessel off location. With the drill pipe suspended from either the upper pipe rams or the middle pipe rams, the drill pipe can be disconnected at that tool joint, pulled and the blind rams closed. Then, the marine riser, kill and choke lines, control hoses and guide lines can all be disconnected. A float valve must be in use in the drill string to permit this operation if the well has surface pressure.

4. *Being able to reestablish the drill vessel on location, monitor and circulate the well prior to reestablishing the drill pipe.* The drill vessel is first returned to the location and the guide lines, marine riser, control hoses and kill and choke lines are reestablished and tested. The choke and kill valves can then be opened and the drill-pipe and annulus pressures observed. The well can be circulated by pumping down the line to the upper side-outlet. When the well has been fully stabilized, the blind rams can be opened and the drill pipe reconnected.

5. *Having alternate methods available with the failure of any one function.* Several specific emergency features have been designed into the system. The dual kill and choke valves provide two means of closing each line at the stack. The lower pipe rams provide an alternate to the middle pipe rams. The Hydril-type preventer provides an alternate to all of the pipe rams. And, as a last resort, the blind-shear rams offer the possibility to cut the pipe and close in the well. The full working pressure of the upper connector provides the possibility to add additional preventers should the need occur.

Note that the arrangement of the preventers and the side-outlets minimizes the number of connections beneath the lowest pipe rams and the wellhead. The lower pipe rams should not be used routinely and should serve as a standby in case of problems with any of the equipment above. The most important action of all is to land the drill pipe as soon as practical to prevent movement of the pipe through the preventer elements.

Testing, inspection, maintenance

Prior to the initial installation of the subsea blowout-preventer stack and while it is still installed on the test stump on the drill vessel, all preventers and kill and choke valves should be functional and pressure tested. After installation, each of the functions should be operated daily and pressure tested weekly and after setting every casing string.

Every time the blowout-preventer stack is pulled and surfaced, the rams should be removed, cleaned, inspected and lubricated. The rubber in the Hydril-type preventer should be visually inspected for wear, tears and gouges. Worn or damaged rams and rubber elements should be replaced.

All bolts on clamps, flanges and other connections should be checked and tightened every time the stack is pulled.

All preventers and the kill choke valves should be operated and pressure tested before re-running to the ocean floor.

On completion of each well, the blowout-preventer stack and the connectors should be carefully inspected for internal wear or other forms of damage.

11

The Control System

✳

The control system must provide individual control for each preventer, valve, regulator, latch, and other subsea function. Also, it must provide power to each rapidly and repeatedly from a completely independent power source. The basic design criteria should include: (1) reliability; (2) simplicity in operation and handling; (3) minimum response time; (4) safety of both the control system and the well; and (5) adequate power-fluid capacity and pressure for each function.

Just as in land drilling, all blowout preventers, valves and other equipment on the subsea blowout-preventer stack are hydraulically operated. Two types of systems for directing the flow of power oil are in common use—the direct system and the indirect. The direct system has individual power-oil lines from the drill vessel to the individual functions on the subsea stack. This is quite similar to the system used on land rigs except that control lines are longer. The indirect system has only one source of power oil to the subsea stack, and flow is distributed to the various functions by pilot valves located on the ocean floor. The pilot valves are actuated from controls on the drill vessel.

The direct system

As mentioned previously, the earliest floating-drilling systems were as closely related to their land counterparts as practical. With the blowout-preventer system located on the ocean floor, the control lines were simply given extra length to reach the individual function. A few extra lines were provided to operate hydraulically some of the additional equipment required on the subsea location; to wit, the connector between the blowout-preventer stack and the wellhead, and the remotely operated kill and choke valves.

These lines were generally ½-in. diameter hoses, reinforced with wire braid, taped together in a bundle, and run to the subsea blowout-preventer stack over the side of the vessel.[1]

The direct system has proven entirely adequate for shallow-water operation, or to a water depth of about 300 ft. It has the principle advantages of being simple and inexpensive, requiring minimum maintenance, and being better known and understood by drilling personnel. All control of the system is maintained onboard the drill vessel. Valves and other equipment can be repaired and pressures minutely controlled on the surface. The hoses are of sufficient diameter that foreign material or other contamination of the power fluid is not critical as to plugging of the lines. In shallow water, response time is adequate.

The direct system has the disadvantages of a bulkier hose bundle than required in the indirect system and of slower response in deeper water. Power fluid is also discharged back to the surface, further increasing friction and response time.

The indirect system

The indirect system and its component parts are unique to floating drilling. It is less understood than the direct sys-

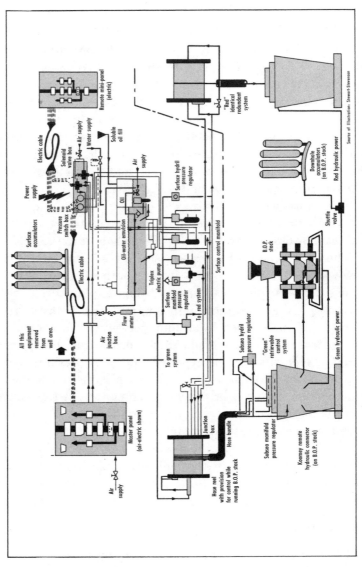

FIG. 11.1   Typical Air-Hydraulic Subsea B.O.P. Control System
(Indirect)

110

tem, and the remainder of this chapter will be devoted to it.

Principal components of the indirect system are: (1) the accumulator unit, or power-fluid supply; (2) the control panel that controls the individual pilot valves; (3) the reels and hoses that transport the fluid to the ocean floor; (4) the control pods; and (5) the distribution system mounted on the subsea blowout-preventer stack.

The indirect system is more complicated, more expensive and requires more maintenance by specialized personnel than does the direct system. It does have the distinct advantages of quicker response in the deep water and of a lighter-weight, easier-to-handle hose system connecting the surface with the subsea stack.

The hydraulic fluid

The operating fluid used in the control system is a soluble-oil and fresh-water emulsion. Several commercial soluble oils have been proven suitable and are generally mixed in a ratio of about 1 part of soluble oil to 10 parts of fresh water. Laboratory tests of the mixture should be made prior to usage to determine the stability of the emulsion.

For operations in Arctic regions or where the hydraulic fluids may be subjected to freezing temperatures, chemically pure ethylene glycol can be used as an antifreeze agent in the water phase. Some of the soluble oil-water emulsions are not stable in the presence of glycol. Others are not stable at below-freezing temperatures. Simulated conditions in the laboratory are essential to evaluating soluble oils intended for use at lower temperatures.

Following are results for soluble oil tested for cold-water service. Water containing 38.5% pure ethylene glycol forms a stable emulsion when mixed in a ratio of 9 parts to 1 part of the soluble oil, even at a temperature of –10°F. Water containing 52.5% pure ethylene glycol and in the same mix ratio forms an emulsion stable down to –40°F.

The accumulator unit

The accumulator unit is generally skid-mounted and contains: (1) a storage compartment for the soluble oil; (2) a controlled-liquid-level, emulsion-storage compartment with automatic additions of water and soluble oil, as needed; (3) a power pump with automatic controls for delivering the hydraulic fluid under pressure; (4) accumulator bottles for storing a ready reserve of hydraulic fluid under pressure; and (5) a manifold of air-operated valves and regulators for distributing the flow of the fluid. Filters are provided in the pump suction to insure clean fluid for the control lines.

Accumulator bottles.

The accumulator bottles make available a quick ready source of power fluid. A simple mechanical device, they store energy in the form of compressed gas. They are precharged with gas at some predetermined pressure, then fluid is injected to compress the gas further. The accumulators used on floating drilling rigs are usually precharged with nitrogen to 1,000 psig. Further compression is achieved by pumping in power fluid to 3,000 psig. The final contents of the bottle by volume are 33.7% gas and 66.3% hydraulic fluid. For calculating the required storage capacity, it can be assumed that roughly 60% of the volume of the accumulators represents usable fluid at 1,000 psi. Depending on the water depth and the individual requirements of the preventers, higher minimum pressures are frequently required. The higher minimums reduce the percent of usable fluid from the accumulator. Boyle's Law

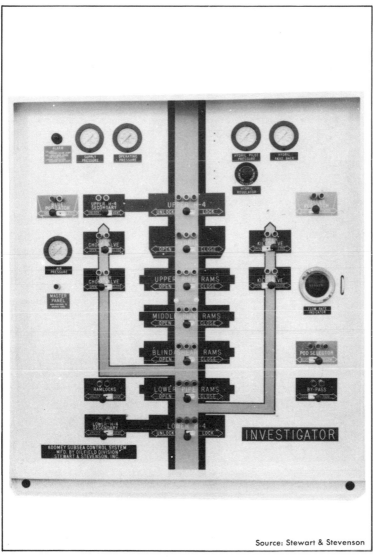

FIG. 11.2 Graphic Air-Operated Master Control Panel With Electric
Read Out

113

FIG. 11.3   Accumulator Unit for Air-Hydraulic Subsea System

$(P_1V_1 = P_2V_2)$ should be sufficiently accurate for this calculation.

As a rule of thumb, fluid-storage capacity should be sufficient to open and close all blowout preventers and still retain a liquid reserve of 25%. Take, for example, a 13⅝-in., 5,000-psi stack with one Hydril GK preventer and four Cameron Type U preventers. Storage capacity is calculated, as follows (Refer to Table 9):

Close Hydril	17.98 gal
Open Hydril	14.62 gal
Close 4 Type U, 4 x 5.8	23.2 gal
Open 4 Type U, 4 x 5.45	21.80 gal
	77.60 gal
25% reserve capacity	19.40 gal
Total storage requirements (usable fluid)	97.00 gal

When considering that only 60% of the volume capacity is usable fluid, then a volume capacity of at least 161.7 gal $(97.00 \div 60\%)$ is required. Cylindrical units of a 10-gal nominal size are the most commonly used. Hence, if each unit has a full 10-gal capacity, 17 units would be required to furnish the full 25% excess capacity. Some units being sold as 10-gal nominal size have a true capacity less than that. This actual capacity should be determined in selecting the number of units needed.

Accumulator bottles are available in cylindrical and spherical shapes and of the nonseparator and separator types. In the nonseparator type, the two fluids (the gas and the hydraulic fluid) are in direct contact. Hence, the hydraulic fluid is subject to absorption of the gas. In the separator type, a bladder, diaphragm, or piston separates the fluids; this is preferred for hydraulic control systems used in floating drilling operations.

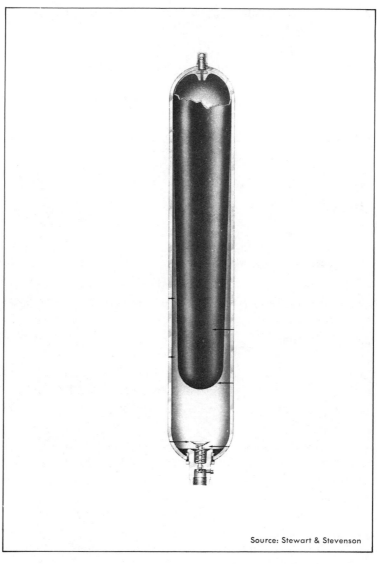

FIG. 11.4   Separator Type Clylindrical Accumulator Bottle

116

TABLE 9
Operating Data On Blowout Preventers

Preventer type, size			Volume requirements, gal.	
			To open	To close
Hydril, GK,	13⅝-in.,	5,000 psi	14.62	17.98
Hydril, GK,	13⅝-in.,	10,000 psi	24.66	34.53
Hydril, GK,	16¾-in.,	5,000 psi	19.91	28.67
Cameron, U,	13⅝-in.,	5,000 psi	5.45	5.8
Cameron, U,	13⅝-in.,	10,000 psi	5.45	5.8
Cameron, U,	16¾-in.,	5,000 psi	4.91	5.29
Shaffer, LWS,	13⅝-in.,	5,000 psi	2.86	3.25
Shaffer, LWS,	16¾-in.,	5,000 psi	5.11	5.60

The control panel

Most control panels have a graphical representation of the subsea blowout-preventer stack to indicate clearly the function of each control. Then, each control's current position is shown by indicator lights. Since these lights are not operated by contact switches but by pressure switches, they give true indication of the pilot pressure being applied in the desired location. All other pertinent pressures are also shown by gauges on the panel.

Meters on the control panel register the volumes of power fluid consumed by the subsea function. This measure further monitors the performance of the equipment.

Typically, a remote-control panel is located at the driller's position and the master panel at another, more removed location on the drill vessel. These panels transmit either air or electric signals to position four-way valves on the accumulator-unit manifold. The valves are then actuated by air pressure, or solenoids, to direct power to the individual pilot lines in the hose bundle connecting to the ocean floor.

The hose bundle

In order to avoid possible leaks and to simplify handling, hose bundles are usually fabricated and supplied as one continuous length. The length of the hose depends upon the water-depth requirements of the rig. In fact, hose length is one of the factors that establishes water-depth capability for a drill vessel.

The control-hose bundle includes one large power-fluid-supply hose (usually 1-in.) and a number of smaller pilot hoses. The pilot hoses connect to pilot valves in the subsea control pod or, in some instances, connect directly to functions. Various combinations of sizes and number of hoses in the bundle are available, depending on the number and type of functions required. The pilot hoses are commonly 1/8- and 3/16-in. diameter; the size affects reaction time and is selected on the basis of the importance of reaction time. The group of hoses are bundled as conduit and covered with a polyurethane protective covering.

Table 10 shows typical hose bundle's number and size of lines and the functions served. Note that the larger 3/16-in. lines are connected to the pilot valves serving preventers and kill and choke valves where reaction time is important. Or, they are connected as direct function lines that do not operate pilot valves. The 1/8-in. lines are used to pilot valves where the time element is not critical or as direct lines where volume requirements are small.

All of the hoses should have a minimum working pressure of 3,000 psi. Commonly, however, hoses with a working pressure of 10,000 psi are used for the small lines. The higher-pressure hoses generally have less expansion due to pressure and, as a result, provide faster reaction time.

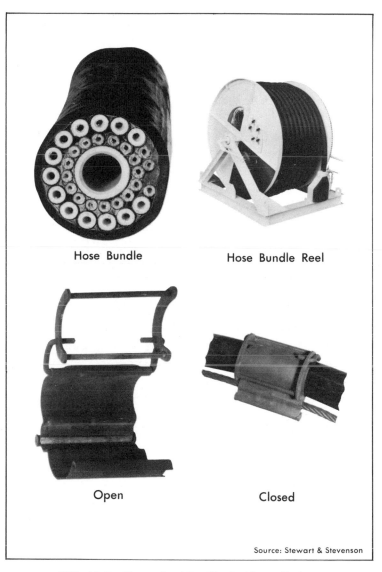

Hose Bundle

Hose Bundle Reel

Open

Closed

Source: Stewart & Stevenson

FIG. 11.5 Elements of the Control-hose System

TABLE 10

Typical Hose Bundle and Functions Served

Subsea function	Number and size of lines		
	1-in.	3/16-in.	1/8-in.
Power-fluid supply	1*		
Common pressure to pilot valves		2*	
Riser connector, primary			2
secondary		1*	
Control pod latch			1
Choke and kill-line connectors			2*
Wellhead connector, primary			2
secondary		1*	
Hydril-type preventer		2	
Hydril pressure regulator		1*	
Hydril pressure monitor			1*
Ram-type preventers (4)		8	
Ram locks			2*
Kill and choke valves		4	
Auxiliary, other, reserve lines		1	2
Total lines	1	20	12

*Directly connected lines.

The hoses are stored and handled on the drill vessel by air-driven hose reels. The reels usually have several live connections that permit operation of certain functions even while the reel is in motion. The live functions will permit the operation of certain latching mechanisms prior to the full establishment of the subsea blowout-preventer stack on the ocean floor. After the hose has been reeled out, and the reel locked into position, all of the additional hoses are connected with a single, quick connect-disconnect junction box.

The hose passes over a sheave located and aligned over the moonpool. The sheave protects the hose as it is run and retrieved and while it is in the operating position. The weight of the hose is supported by clamping the hose to wire cable working on an air tugger.

The control pod

The control pod serves as a lower terminal for the hose bundle and houses the critical moving parts of the subsea hydraulic-control system. The pilot valves and the pressure regulator for the Hydril-type preventer are all contained within the pod.

The pod also serves as a junction box with the subsea blowout-preventer stack. The latching mechanism is remotely controlled from the drill vessel so that, in the event of a malfunction by any of the components within the pod, the entire pod can be recovered and serviced.

The same wire cable that supports the hose bundle is attached to the control pod and is used for running and retrieving the pod.

Different manufacturers' control pods and components perform similar functions although each has individualistic features and design. For simplification, the equipment discussed here is only one of the various designs available, although other designs are equally acceptable.

The pilot valves are three-way and normally installed with the stack function open to exhaust to the ocean floor. When pilot pressure is applied, the valve shifts, closing the exhaust ports and opening power supply oil to actuate the function on the stack. When pilot-valve pressure is released, spring loading in the valve returns it to its original position and vents the pressure of the stack function.

The regulator valves are adjusted from the surface by regulating hydraulic-fluid pressure connected directly to the hydraulic head of the regulator valve in the control pod. Another direct line connected downstream of the regulator valve furnishes to the surface a read-out of the downstream pressure.

The retrievable male section of the control pod is latched into the subsea female section by means of hydraulically operated latching dogs. As the male section is lowered into position, hydraulic fluid is circulated through the control line and vented to the ocean floor. As the male section moves into final position, the exhaust of the control line is automatically sealed. The fluid pressure then increases, and the latching dogs are actuated. The pod is recovered by releasing the pressure and exerting a vertical pull to retract the latches.

Redundancy

Most floating-drilling systems are equipped with dual hydraulic-control systems between the drill vessel and the subsea blowout-preventer stack. The system is, thus, completely redundant. Two identical hose bundles and control pods are installed. By means of control from the drill vessel, either may be used to perform any subsea function. If either hose bundle is damaged, or even severed, the fluid power may be switched to the alternate hose and full control of the subsea equipment maintained. If one of the components of the control pod develops a leak or malfunctions, the same flexibility exists and the operation may be switched to the alternate pod.

Control system on the subsea blowout-preventer stack

The hydraulic system on the subsea blowout-preventer stack is relatively simple. All pilot valves and the control of the flow of the hydraulic fluid are located within the control pods. Only the piping from the control pods to the individual functions is needed on the stack itself. With the use of two control pods, however, some type of check-valve arrangement is necessary to isolate any problems on one of the control systems. The junction of the two sources of power oil for the individual stack functions is made with a simple tee-junction block. The junction block contains a shuttle valve which shifts according to the direction of flow to the function. The shuttle valves have proven virtually trouble and maintenance-free.

Most subsea blowout-preventer stacks are being equipped with accumulator bottles of 50 to 100-gal capacity and mounted directly on the stacks. Original purpose of installing the bottles was: (1) to provide a source of hydraulic power close to the function; and (2) to decrease the time required for furnishing power oil from the surface. From a practical standpoint, the reaction-time advantage is insignificant when power fluid is supplied through hoses of 1-in. diameter. A side benefit of having a storage of power fluid on the subsea stack has developed, however. This arrangement provides a source of power to the subsea functions even if the vessel is swept completely off location and all control hoses and other connections with the ocean floor are lost. The power from these subsea accumulator bottles can operate either automatically or with simple, one- or two-function acoustic controls.

Automatic closing for the Hydril-type preventer

At least two floating drilling vessels working for an operator in the Santa Barbara Channel, California, have been equipped with automatic closing devices for the Hydril-type preventers.[2]

These systems are designed so that, if control from the surface is lost, they will: (1) shut-off the connection with the surface to retain the charge in the subsea accumulators; and (2) shift the pressure automatically to the close function of the preventer. If this occurs when pipe is out of the hole, the preventer then closes in the well. If it occurs when drill pipe is extending through the preventer, it then closes around the pipe. With a float valve installed in the drill string, the well is also closed in until the vessel can be reestablished.

The automatic system is designed so that it can be, by choice, rendered inoperative if some particular operations make this desirable.

Operation of the indirect system

Dual control systems are provided primarily to offer dual opportunities to control wells during emergencies. They are *not* for the purpose of being able to continue operations after the malfunction of any one part of the system. When any one part of the control system becomes inoperative, operations should be delayed until that function has been repaired or replaced.

The use of the two control systems should be alternated weekly to test the functional ability of both systems and to reduce the deterioration common in unused equipment.

Field experience has shown relatively few operating problems with indirect control systems. Some hose leakage has been reported and, occasionally, pilot valves have required repair. The practicality of the recovery feature of the control pods has been proven, even in water depths to 1,300 ft.[1]

Very little information has been published on the reaction time of subsea equipment relative to the variables involved.

In general, subsea blowout preventers, as presently installed, require 15 to 80 sec. to operate depending on hose size, water depth, volume requirements, mud weight, and control equipment. Hose size and length is the most important variable for any given volume requirement. For example, a 13⅝-in. Hydril-type preventer requires about 40-45 sec. with a 1/8-in. pilot-line hose operating in 1,500 ft. of water. In only 500 ft. of water, this time reduces to about 30-35 sec. By increasing hose size to 3/16-in., reaction time can be reduced to about 25 sec. for either water depth. Of course, ram-type preventers require less time to operate because of their lower volume requirements. Conversely, all of the 16¾-in. equipment requires greater reaction time because of the larger volumes.

Testing, inspection, maintenance.

Of major importance to minimizing control-system problems is the maintenance of a clean hydraulic fluid, free of all trash or foreign material and salt water. Filters in the accumulator units should be checked frequently and maintained in first-class condition.

Salt water can form tight, pasty emulsions with some of the soluble oils, and care should be exercised to avoid salt-water contamination. If contamination does occur, the system should be dumped and all lines flushed and cleaned before replacing the fluid.

Operating pressures for the various preventers should be set for the operating water depth in accordance with the manufacturer's recommendations.

Spare parts maintained on the drill vessel should include replacement pilot valves, regulators, resilient seals for the control pods, and extra hose and fittings for the hose bundle.

Alternate, and future control systems

Combination electric-hydraulic systems have been available for several years but, to date, have not been used extensively. In the combination system, the pilot valves are actuated by electric power and have the distinct advantage of very low reaction time. Perhaps the major reason that the system has not been more widely accepted is its subsea electrical connection that originally was a source of problems. Users of the system contend that this is no longer a problem.

A fairly new system now available uses the more conventional control pod with electric actuators for the pilot valves. With this system, the only subsea connection is the standard, proven hydraulic connection; no electrical connections are made underwater. Reaction time is greatly reduced and the hydraulic hose bundle is replaced by a more compact electric cable. In years to come, this should become popular for waters deeper than 1,000-ft.

Several companies are currently testing acoustically controlled equipment. Ultimately, this could eliminate all hardware between the surface and the subsea package. It assumes a practical method can be developed to store the required electric energy on the subsea package for actuating the valves and repressuring the hydraulic power fluid. Without the subsea power supply, acoustic controls for the entire system offer little advantage.

REFERENCES

1. Bezner, H. P.: "The Evolution of Blowout Preventer Control Systems and Riser Tensioning Systems". Spring Meeting of Pacific Coast District, Division of Production, API, Paper No. 801-45E, May 1969.

2. Harris, L. M., and Ilfrey, W. T.: "Drilling in 1,300 ft. of Water—Santa Barbara Channel, California." First Offshore Technology Conference, OTC-1018, May 1969.

12

The Marine Riser System

Marine drilling risers are used to provide a return fluid-flow path between the well bore and the drill vessel and to guide the drill string to the wellhead on the ocean floor. No pressure integrity is required of the marine riser other than for the differential in hydrostatic pressure between the drilling fluid and the sea water. The marine riser, however, must withstand the lateral forces of the waves, currents, and vessel displacement. It must also withstand the axial loads imposed by the buoyancy weight of the drilling mud, drill pipe, and the marine riser itself. Then, too, with a tensioned riser system, the riser must withstand the axial tension imposed on the surface.

The marine-riser system, as discussed in this chapter, includes (from bottom to top): the marine-riser connector; a lower flexible joint; individual riser joints with their connectors; and a telescopic joint. In some riser systems a second flexible joint is used between the telescopic joint and the top riser joint.

Theory and procedure

Fischer and Ludwig[1] describe the behavior of the riser "not as a riser-buckling problem but as a beam-deflection problem for which lateral deflection can be enormously increased by axial compression or, on the contrary, substantially decreased by axial tension". The axial force can vary along the length from tensile in the top to compressive in the bottom. Further, deflection must be determined with variable axial tension. The amount of deflection, rotation, and bending is solved by an ordinary linear fourth-order differential equation.

Dynamic effects are not considered in this general solution. Approximate methods of calculation indicate that, if static design requirements are met, the added cyclic deflections and stresses due to wave forces will not be significant in currently used riser sizes and water depths.

Tidwell and Ilfrey[2] using the same approach as Fischer and Ludwig presented analysis techniques and resulting stress concentrations. Their work reports the studies by Humble Oil for the design of a single-wall tensioned riser for water depths to 1,500 ft. in the Santa Barbara Channel of California. They, together with Harris and Ilfrey[3], also published details on design criteria and performance of this installation.

In the tensioned-riser concept, loads tending to bend the riser are resisted by axial tension applied at the top of the riser. Potential bending loads are exerted by wave and current forces. Any tendency of the riser to bend under these loads is enhanced by the weight of the riser and by the drilling fluid and drill pipe in the riser. Some opposition to bending is supplied by the seawater surrounding the riser. Increasing offset of the top of the riser with respect to the subsea well tends to enhance further the bending effects of the riser, the drill pipe and the drilling fluid.

Axial tension applied at the top of the riser tends to reduce bending, thus reducing the stress in the riser due to bending. However, as tension increases, axial stress in the riser tends

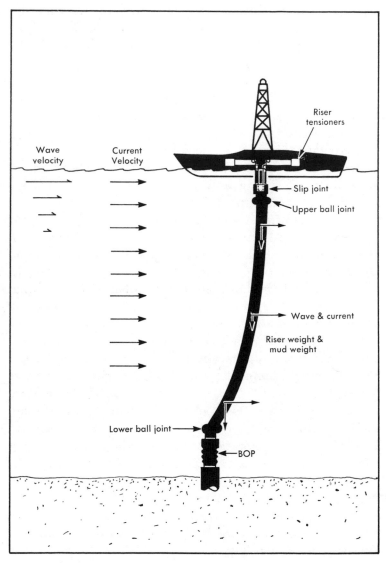

FIG.12.1 Riser System for Floating Drilling Rig

130

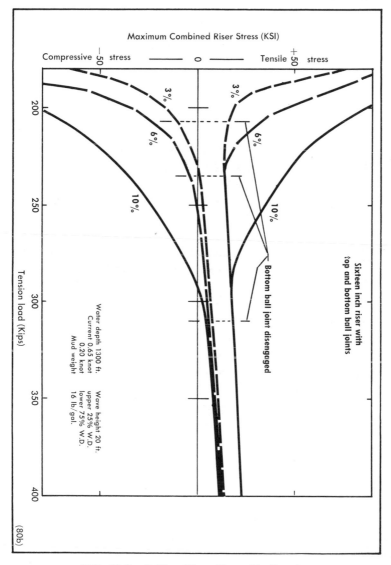

FIG. 12.2 Drilling Riser Stress Vs. Tension

131

to increase. For any combination of oceanographic conditions, mud weight, riser weight, and vessel offset, there is an axial tension for which the sum of axial and bending stresses in the riser is a minimum.

Tidwell and Ilfrey used a computer program based on a finite-difference solution of the differential equation describing beam-column behavior under static loading with small deflections. Their work yielded computer-generated plots of maximum total stress in the riser versus total applied tension. Thus, the location in the riser of the point of maximum stress was determined.

This work shows that maximum stress reaches a low point for a value of tension near that which disengages the flexible-joint stop, Fig. 12.2. Further increases in tension have only a small effect on maximum stress. Moreover, maximum stress is insensitive to vessel offset for the higher valves of tension.

The drilling riser is subjected to dynamic loadings so that stress varies with time. The design of the riser system must, therefore, consider the possibility of riser failure from fatigue as well as from overstress. In addition to cyclic stresses due to wave and current loading (vortex shedding), applied tension may also vary, resulting in a stress variation. Fatigue considerations, taken with the findings from the static study, suggest that the tension setting minimize the possibility of riser failure due to either overstress or fatigue. The tension setting should produce a stress level such that large variations in either vessel horizontal displacement or riser tension have only small effect on maximum stress variation.

Riser-pipe selection

Riser-pipe size is determined by the size of the blowout-preventer stack and the wellhead, with allowance for clearance in running drilling assemblies and casing. For a 13 ⅝-in wellhead and blowout-preventer stack, a 16-in. OD marine riser with a 0.375- or 0.438-in. wall thickness is ideal. For the larger 16¾-in. subsea assembly, an 18⅝-in. marine riser with a 0.500-in. wall thickness is recommended. Selection of the riser-pipe steel is critical for a long-life, trouble-free operation. The steel must have a minimum yield strength well in excess of the maximum stress generated. Yet, it must have good fatigue characteristics and be weldable. Humble has reported good experience with X-52 steel in deepwater operations in the Santa Barbara Channel, and X-52 pipe should be suitable for most operations. However, the HY-80 (A 543, Class I) steel used in submarine construction appears superior to X-52 in strength, toughness and impact resistance. Hence, it is recommended for marine risers, particularly for deep-water, critical operations.

Integral choke and kill lines

Three types of marine-riser systems are currently in use— the integral, the track-type, and the funnel-type. The three differ in the manner that the kill and choke lines are installed. The integral marine-riser system has the choke and kill lines installed on the riser joints so that they are simultaneously stabbed and made up during the stab and make up of the marine-riser connector. This eliminates the handling, rigging, make up and running time normally associated with the choke and kill lines. It is the preferred marine-riser system of the three. The integral marine-riser system can be installed and retrieved in 8 to 24 hours less time than either of the other two systems, depending on water depth.

Integral marine-riser connectors

Marine-riser connectors have been designed to minimize installation time of the riser system. They should provide a joint capable of withstanding the high tension loads applied in tensioned-riser systems and the varying stress loads imposed by wind, wave, and current forces and vessel movement. Quick-stab mandrel-and-box type connections are used with mechanical locking devices to secure the joint. O-ring type seals have been proven adequate and are simple and inexpensive to maintain. Alignment pins are necessary in either the track or funnel types of marine-riser systems but not in the integral riser systems. In this latter system, the kill and choke-line connectors provide alignment and prohibit joint rotation.

The kill and choke-line connectors have chevron-type packing and do not require mechanical latching since the joints are held in position by the riser-pipe connection.

Several points are worth serious consideration in selecting the connector in the integral-riser system. The connectors should be of forged steel having the desired strength and weldability; welding is necessary to attach the connectors to the riser pipe. Strength tests of the connector should prove it adequate to withstand bending (under tensile load) in excess of the riser-pipe strength and tensile loading at least twice the maximum anticipated operating loads. The connector should be a smooth tapered design to avoid stress concentrations due to section change in the weld area. The connector should be of minimum weight consistent with the strength required. The joint when made up and tested under load should have no lateral, vertical, or rotational movement. Three connections must seal simultaneously—the riser pipe,

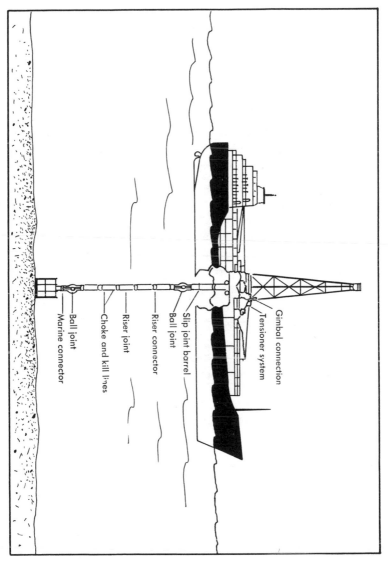

FIG. 12.3 Marine Riser System Components

the kill line, and the choke line. Hence, some provision must be made for flotation or lateral movement of the kill and choke-line connectors. Besides meeting all the above specifications, the design should also be efficient for handling, running and pulling, and require minimum maintenance.

The marine-riser flexible joint

Flexible joints are used in the marine-riser system to minimize the bending moments and stress concentrations. They are commonly placed in the bottom of the system just above the blowout-preventer stack. In deepwater operations or in unusually severe sea conditions, a flexible joint should also be provided in the top of the system just below the telescopic joint. This reduces stress concentrations created by wave forces in this zone and by the change in section between the telescopic joint and the top marine riser joint. Tidwell and Ilfrey presented data supporting this top flexible member.

Flexible joints should be selected on the basis of: (1) providing adequate angle of flexure for the total floating-drilling-system design, usually about 10°; (2) having sufficient strength for the tension to be applied; and (3) rotating with minimum resistance while under the full anticipated tension load.

The marine-riser telescopic joint

The telescopic joint serves as a connection between the marine riser and the drilling vessel, compensating for the vertical movement of the vessel. In operating position, the upper member (or inner barrel) is connected to, and moves with, the drill vessel. The lower member (or outer barrel) is then an integral part of the marine riser and remains stationary in respect to the ocean floor.

handling and in measurements. Adequate pup joints are necessary for final space out. Where 40-ft riser joints are selected, one pup joint each of 20-ft, 10-ft and 5-ft are needed.

Inspection and maintenance

When removed from service, the marine-riser joints should be cleaned, visually inspected for wear or damage, have seals replaced as necessary, and lubricated. The weld areas of the riser joints should be sand-blasted and magnafluxed periodically. The frequency of this type of inspection depends on the severity of the service and the water depth, but in no case should exceed one year. Following sand-blasting and inspection, the joints should be repainted to protect against corrosion.

Special equipment for handling the riser joints has proven beneficial, both to protect the equipment and to improve efficiency in handling. A flare-end guide tube to guide the riser connectors through the rotary during pulling operations reduces damage to the joints and speeds handling. A joint-laydown trough installed in the V-door during running and pulling of the riser has also helped to reduce damage and to increase handling efficiency. A special racking area and racking beams designed for the marine-riser pipe protect the joints while stored on the vessel.

REFERENCES

1. Fischer, W., and Ludwig, M.: "Design of Floating Vessel Drilling Riser". Society of Petroleum Engineers of AIME, SPE-1220, October 1965.

2. Tidwell, D. R. and Ilfrey, W. T.: "Development in Marine Drilling Riser Technology". ASME Petroleum Mechanical Engineers Conference, 69-Pet-14, Sept. 1969.

3. Harris, L. M. and Ilfrey, W. T.: "Drilling in 1,300 Feet of Water—Santa Barbara Channel, California". OTC 1018, May 1969.

13

The Marine-Riser Tensioning System

✳

The importance of adequate tensioning to the life of the marine riser has been discussed. The tensioning system has the dual purpose of compensating for the motion of the vessel while maintaining the tension constant.

Marine-riser tensioning can be provided either by a deadweight system or by the use of pneumatic tensioner cylinders.

The deadweight system

The simplest floating drilling system used in shallow water utilizes a free-standing marine riser and requires no tensioning. For relatively calm seas, the free-standing concept is good up to about 200 ft of water depth. With the addition of buoyancy modules on the riser pipe, this free-standing concept could possibly be extended into deeper water.

For water depths that require tensioning marine risers up to about 100,000 lb., the deadweight system has proven satisfactory and generally the simplest. Counterweights are suspended by wire rope through a system of sheaves attached to the lower half of the telescopic joint. The counterweights are free to rise and fall as the vessel moves with the sea.

Hence, the system fulfills both the requirements of desired tension and motion compensation.[1]

As higher tensions are required, the disadvantages of the deadweight counterbalance system become apparent. Tensioning the marine riser adds to the vessel load. Counterbalancing the tensioning with weights doubles the load of tensioning. This is particularly disadvantageous on semisubmersibles where deck loading is already limited.

To vary the tension in the marine riser requires varying the amount of counterweight. Most deadweight-counterbalance systems do provide means of adding or removing weight sections. These sections, however, are difficult to handle and are almost inaccessible. The design must protect the vessel in the event of a tension-line failure. Falling weights could penetrate hull sections and, hence, must fall free of critical vessel compartments.

The pneumatic tensioning system

Most floating drilling vessels are now equipped with an arrangement of pneumatic cylinders. These are designed to provide pneumatically controlled tension for each of several lines attached to the marine riser. Two equipment manufacturers now market units suitable for this service.

One common riser-tensioning unit has a 12-in. diameter piston acting within a cylinder, with dual 42-in. diameter sheaves mounted on opposite ends of the unit. The unit has a 10-ft piston stroke but, through sheaving of the lines, provides compensation for 40 ft of vessel movement. Air pressure up to 2,200 psi acts on the pistons to exert line tensions up to 60,000 lb. The rod end of the cylinders are filled with a

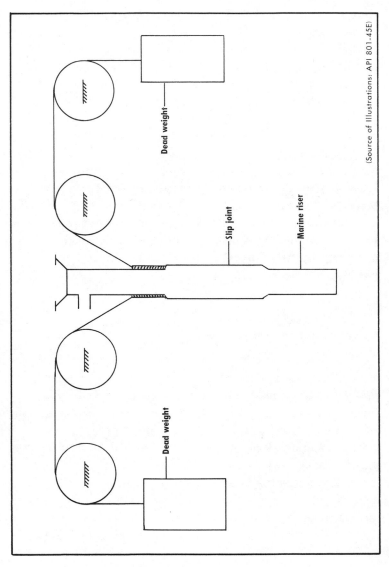

(Source of Illustrations: API 801-45E)

FIG. 13.1 Dead Weight Riser Tensioning System

142

hydraulic fluid to perform the dual purpose of dampening the action of the piston and lubricating the packing.

The principal components of the unit then are: the cylinder with the air-balanced piston and sheaves; an air-oil reservoir to supply hydraulic fluid to the piston end of the cylinder; and air accumulators that supply air to the high-pressure side of the piston and regulate the tension.

Most drilling vessels today are equipped with four of these units, or have a rated line-tensioning capacity of 240,000 lb. A few vessels are equipped with six, or a rated line-tensioning capacity of 360,000 lb. From a practical standpoint, the units probably should not be used on a continuous basis to supply more than about 80% of rated load, or 48,000 lb per unit. This would mean that the 4-unit and 6-unit installations actually have a continuous line-tensioning capability of 192,000 and 288,000 lb, respectively. Some problems have been reported with piston leakage when operating at higher than the 80% capacity loads.[2] New designs of pneumatic tensioners incorporate an oil bath on each side of the piston that may increase the life of the packing on the piston and extend its performance capacity.

The pneumatic tensioning system has proven helpful in landing and latching the subsea blowout-preventer package on the subsea wellhead. When the preventer stack has been run on the riser pipe, the tensioners are connected to the telescopic joint just before landing the stack. The weight of the riser and stack are then suspended on the tensioners and, gradually, lowered into place. After latching, the tensioners should be brought up to the desired tension in increments of about 40,000 lb and tensions verified with portable cable-tension indicators. Pressure gauges normally used to monitor the tension on the riser lines should be calibrated quarterly to insure accuracy. NOTE: Pressure gauges measure tension in the lines. Tension for the riser, however, is a function of the line tension and the angle between the lines and the riser.

Riser-tension-line experience

The wire rope used in marine-riser tensioning systems is subject to extensive cycling. Hence, most deterioration is a result of fatigue stressing. Failures reported to date indicate that failures have occurred after about 12.4×10^6 ton-cycles of operation. This is subject, of course, to the environmental conditions of the wire rope (whether subjected to rig-floor wash water, etc.), lubrication, and the type of wire rope used. Until additional performance data becomes available, it is recommended that a 6X37 or 6X43 wire rope be used for this service and retired after 7.0×10^6 ton-cycles. Records need to be maintained on the accumulated service. The service can be accurately calculated using a counter on one of the sheaves or estimated from the normal period of the seas in the area.

Inspecting the lines while in service is normally difficult in that the area most subject to wear is hidden on the sheave. With evidence of breakage of 10% of the wires in one lay of the wire rope, the line should be retired from service immediately. With evidence of six or seven of the wires in any one strand broken close together, the line should be retired from service immediately. Frequent use of a lubricant prolongs the life of the wire rope.

When the wire rope has accumulated sufficient service to warrant replacement, it should be replaced in its entirety. Often attempts are made to "slip" the line to move the wear spots, as is commonly effective with drilling lines. Because of the number of sheaves and the short distance between

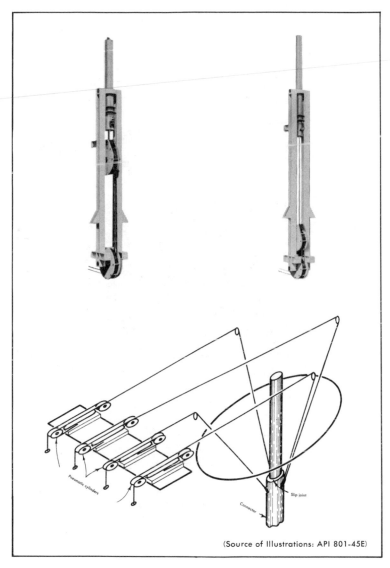

(Source of Illustrations: API 801-45E)

FIG. 13.2 Pneumatic Tensioner System for Marine Conductor

145

FIG. 13.3 Pneumatic Tensioners on Floating Rig

146

them, it is not practical to "slip" the tensioning lines. This is apparent when the line is diagrammed as a straight line on paper and the travel areas are marked.

REFERENCES

1. Bezner, H. P.: "The Evolution of Blowout Preventer Control Systems and Riser Tensioning Systems". Pacific Coast District, Division of Production, American Petroleum Institute, Los Angeles, 801-45E, May 1969.

2. Harris, L. M., and Ilfrey, W. T.: "Drilling in 1,300 feet of Water-Santa Barbara Channel, California". First Offshore Technology Conference, Houston, Texas, OTC 1018, May 1969.

14

The Guide-Line System

✳

Prior to installation of the marine-riser system, all equipment, tools, drill bits, etc. are guided between the drill vessel and the ocean floor by means of the guide-line system. Even after installation of the marine-riser system, the guide lines are sometimes used for running the television camera for subsea inspections.

Too little attention to the design of the guide-line system frequently has created unnecessary problems and caused loss of rig time.

Lateral loads are imposed on the guide lines by vessel offset and by current forces during the time the subsea equipment is lowered to the ocean floor. The effect of these forces on the stress in the lines can be minimized by sufficient tension and by care in the design of the guide arms and guide posts.

Guide-line tensioning

The tensioning system for the guide lines has the same basic requirements as for the marine-riser system. That is, the tensioning system must provide the needed tension to the lines and compensate for vessel motion. Just as with marine-riser tensioning, weight counterbalances and pneumatic tensioning rams have proved satisfacotry.

Experience has shown that ¾-in. wire rope is adequate in strength for guide-line systems working in water depths to 1,500 ft. In fact, water depth has little effect on the design of the guide-line system, except for the length of the lines. Length of the line does affect the working load of the line because of line weight. This is insignificant in the guideline system, however, and more than offset by the reduced currents found near the ocean floor in the deeper water.

Experience has also shown that even the heaviest loads can be efficiently guided to the subsea base with surface line tensions in the magnitude of 8,000-10,000 lb. Tensions of this magnitude are required only while the object is being lowered into position, or recovered. At all other times, tension can be reduced to just sufficient to hold the weight of the line and prevent tangling, or about 2,000 lb. Reducing the tension during the periods of non-use, of course, prolongs the life of the wire.

Primarily because of the variations in tension requirements for the lines, the pneumatic tensioning system is preferred over counter-weights. Pneumatic rams with a 6-in. diameter ram are capable of a line pull up to 14,000 lb and have performed extremely satisfactorily in this service. One ram is installed on each of the four guidelines. Like the riser tensioners, the rams are sheaved so that the 7½-ft stroke of the ram permits up to a 30-ft motion of the vessel.

In the past, many guide-line systems have been tensioned by air winches, or air tuggers. Most of these systems proved inadequate and are obsolete.

Guide-arm funnels and guide-post design

Wire rope passing through guide-arm funnels is subject to the same stresses as when working over sheaves, if the lines and load are not perfectly aligned. The interior of the funnel, therefore, should be designed with a radius of curvature selected by the same standards used in selecting sheaves. A radius of curvature of 12-in. is the equivalent of diameter-to-wire-rope ratio of 32:1 for ¾-in. wire rope and is recommended for use in funnel design. Sharp corners are common in many funnel designs but should be avoided as they create unnecessary high stresses in the guide lines.

Most breakage of guide lines occurs at the guide post on the subsea guide structure. This is directly attributable to inadequate tensioning. A contributing factor can be the internal cutting edges of the guide post acting on the line. As suggested for the guide-arm funnels, the guide posts should be given an internal radius of curvature to remove this point of stress concentration.

Inspection, maintenance, replacement of guide lines

Guide lines normally require replacement because of cuts or kinks in the wire rope, rather than because of wear or fatigue stressing. For this reason the more rugged ¾-in. 6x19 Seale wire ropes are recommended. Galvanizing is usually justified to extend the life of the line.

Visual inspection is about the only practical means of determining replacement requirements. Lines with excessive kinks or damaged strands should be replaced, as necessary. Individual strand breakage of 10% of the wires in any one lay is also sufficient justification for replacement.

15

Support and Auxiliary Systems

Vessel-position reference systems

Of all the pieces and parts that combine to form the total floating-drilling system, few play a more significant role in safe, efficient deep-water operations than the device that monitors the position of the vessel relative to the subsea wellhead. Accurate information on vessel position permits accurate alignment. Accurate alignment minimizes riser stresses, facilitates the latching and unlatching of subsea components, reduces running time for tools and, most important, eliminates unnecessary wear of the subsea equipment.

Two vessel-position reference systems are in common use—the taut-line system and the acoustic system. The taut-line system uses a taut steel line stretched from the vessel to an anchoring point on the ocean floor. It may be a guide line or an independent line with an attached anchor weight. An air winch or counterweight system on the vessel maintains constant tension on the line. A dual-axis inclinometer measures the slope of the taut line at the vessel. The taut line is assumed to be straight from the vessel to the ocean floor, so its slope indicates vessel displacement. However, as water depths increase, drag forces due to current can distort the taut line appreciably, so the assumption of a straight line may be invalid.

Adams[1] developed methods for evaluating the accuracy of the taut-line system for a range of water depths, current velocities, and equipment parameters. He also devised methods to reduce the error in the taut-line systems. His study concluded that under realizable conditions the taut-line system is subject to appreciable error due to current drag and weight forces on the cable. His study further concluded, however, that the system error can frequently be limited to a selected value by appropriate selection of tension, cable size, and anchor weight, as shown in Table 11.

To realize the tensions shown necessary by Adams create problems in the field. Too often, because of these problems, lesser tensions are used and inaccurate measurements are tolerated. For example, where guidelines are used as the taut line, tensions of more than 10,000 lb are required to maintain an accuracy of 1% in water depths below 1,000 ft and with currents of one knot. Tensions of this magnitude significantly increase wear of the line and endanger the integrity of the guidance system. When a separate, independent anchor and line are set, the mass of the anchor required for adequate tension generally precludes recovery of the anchor without exceeding the ultimate strength of the taut line.

The acoustic position-reference system operates without a mechanical link to the ocean floor. A beacon is established on the ocean floor to transmit an acoustic signal. The beacon may be dropped anywhere in the vicinity of the subsea drill site or may be installed directly on the subsea blowout-preventer stack. If it is dropped separately, it has the advantage of providing position information from the start of operations. And, it can remain in place even though the blowout-preventer stack is recovered and the vessel moved off location. It has the disadvantage of requiring predetermined arrangements for the recovery of the beacon. If the beacon is mounted on the blowout-preventer stack, it has the disadvantage of monitoring position only after the stack is in place. And position information is interrupted whenever the stack is recovered to the surface. The installation on the stack does simplify final recovery of the beacon.

TABLE 11
Taut-Line Position Reference System
Error Calculations For Uniform Current[1]

Steel line size, in.	Water depth, ft	Current, knots	Tension, lb	Error, %
⅛	1,000	0.90	153	11.07
			755	1.95
			6,810	0.21
⅛	1,500	1.50	1,000	6.17
			3,000	2.01
			5,000	1.20
⅜	500	0.90	1,000	2.24
			3,000	0.70
			5,000	0.42
⅜	1,000	1.50	1,000	13.85
			3,000	4.02
			5,000	2.36
⅜	2,000	1.50	2,180	12.47
			2,750	9.49
			3,580	7.04
			4,520	5.45
			7,130	3.36
⅝	1,500	1.50	2,400	17.31
			3,000	12.58
			4,000	8.69
			6,850	4.64
⅝	2,000	0.90	3,000	6.76
			5,000	3.38

The signal transmitted from the beacon is received by hydrophones mounted below the hull of the drill vessel. Three hydrophones are required to fix the position but, commonly, four are installed to provide redundancy. Signals are fed from the hydrophones to the vertical-reference unit where compensation is made for vessel roll and pitch. Offset is then converted into x and y position coordinates. The coordinates can be recorded or projected onto a position display unit. The display shows the reference point as a bright dot on a large reference screen, the center of which represents the vessel.

Acoustic position-reference devices are available with an accuracy to 0.5% of water depth in water to 1,500 ft. Less accurate and less expensive units, to 1% of water depth, are adequate for most floating drilling operations.

Acoustic position-reference systems offer several advantages and more flexibility than other systems. Since there is no mechanical linkage between the vessel and the ocean floor, even sudden or unprepared departures of the drill vessel should not disturb the ocean-floor beacon. Later, the beacon can assist in reestablishing the drill vessel on location. Accuracy of the unit is not affected by current or other drag forces. Dual temporary beacons can be mounted on running tools to monitor any rotational motion of the tool between the vessel and the ocean floor; this feature is particularly useful in establishing the temporary guide base. Outputs can be taken from the vertical-reference unit to record pitch and roll information as these data are continually monitored. The hydrophones and certain other components of the system can be used for other acoustic functions, such as acoustic angle indicators mounted above flexible joints.

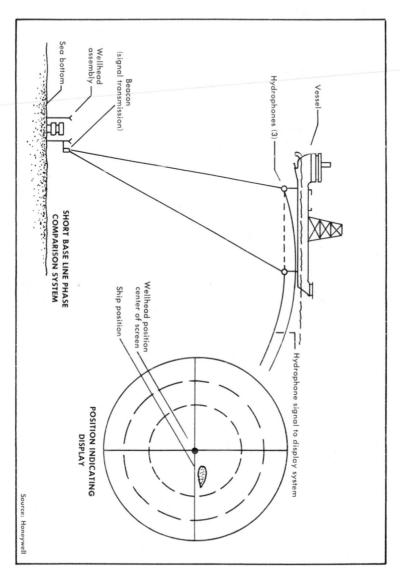

FIG. 15.1 Vessel Position Indicating System

155

Angle-azimuth indicators for marine-riser flexible joints

An angle-azimuth indicator mounted on the lower marine-riser joint just above the flexible joint provides the most positive, practical, and economical means to assure the structural integrity of a marine-riser system.[2] Overstressing of the lower riser joint can be prevented by monitoring the flexible-joint angle and by correcting any vessel displacement or inadequate riser tensioning. Other useful benefits of maintaining a small angle in the flexible joint include: (1) less drill-pipe fatigue and wear; (2) less wear and damage to the subsea blowout-preventer stack and wellhead equipment; and (3) greater ease in running tools and drilling assemblies. Knowledge of the flexible-joint position also facilitates disconnecting the upper connector between the marine riser and the blowout-preventer stack.

Two types of angle indicators are currently in use on floating drilling vessels—an acoustic system and an electric-cable system. The acoustic system is preferred because of the inherent troubles with electric cables connecting the ocean floor with the drill vessel. Also, additional rig time is required for running and pulling the marine-riser system with electric cables attached.

The acoustic system consists of three basic components: an underwater sensor and electronics unit; a hydrophone; and a shipboard electronic processing and display unit. The underwater unit mounted on the riser senses the angle and azimuth and transmits the signal to a hydrophone mounted below the hull of the drill vessel. The hydrophone converts the acoustic pulses which are used as input signals to the shipboard electronic processing and display unit. The angle-azimuth information is then displayed in degrees on a cathode ray tube as with the acoustic position-reference system.

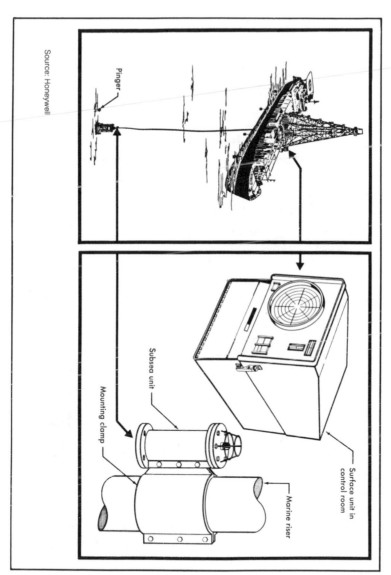

Pinger

Subsea unit

Mounting clamp

Surface unit in
control room

Marine riser

FIG. 15.2 Riser Angle Indicator

Acoustic angle-azimuth indicators are available to operate in water depths to 2,000 ft with an accuracy within 0.5° over a range of 10° from vertical and with a battery life of 100 days. Batteries are rechargeable on location.

The angle-azimuth indicator is needed for full reliability in floating-drilling systems, especially in deeper water. It should not be used, however, in lieu of a position-reference device since it supplies information only on the angle and direction of the bottom riser joint. The magnitude of the angle can be varied by mud weight, by current or other drag forces, or by riser tension even when the drill vessel is perfectly aligned over the hole. Attempts to reduce this angle by only repositioning the vessel can create conditions even more severe for the riser system.

Television

Underwater television cameras have proven practically indispensable for efficient floating drilling operations. In fact, many rigs operating outside U.S. domestic waters maintain a spare camera and television cable to insure quick replacement of troublesome equipment should electronics specialists not be readily available.

Most vessels use a light-weight, tubular constructed frame for lowering the television camera to the ocean floor. The frame is designed to ride on the guidelines and is raised and lowered by an electric cable on a powered winch. Incandescent lights are mounted with the camera on the frame, and provision is made for the control from the surface of the pan, tilt, and focus of the camera.

Television is routinely used to observe the landing of subsea equipment, tool operations and cementing. Frequently, it has served to pinpoint leaks in control systems or other troubles with equipment. The successful use of television for floating drilling in water to 1,300 ft has been reported.[3] It has been used to a limited extent with video-tape recording in supervising and documenting offshore operations and surveys and probably will be used more frequently in this type of work in the future.[4]

Divers and diving equipment

Current equipment and technology make it possible to design floating-drilling systems that can be operated totally independent of the use of divers, submarines, or any manned underwater equipment. Redundancy in equipment and careful forethought to the selection of drilling programs and practices are essential to the fulfillment of this goal. Occasionally, however, divers can be the more economical solution to underwater problems.

Divers have proven their ability to work with a reasonable degree of efficiency in water depths to 840 ft. They have simulated dives to 1,700 ft successfully.[5] In 1970 a French company, Compagnie Maritime d'Expertises (COMEX), conducted actual working dives in 840 ft of water. The divers performed a total of 35 hours work including: (1) connecting and disconnecting flanges with impact wrenches; (2) underwater oxy-arc cutting of a 6-in. pipe; (3) welding in a T-junction section; and (4) assembling a template section requiring dexterity. The crew, operating from a chamber at 660 ft, spent over three hours of continuous immersion at 840 ft in the final working dive.

Instrumentation, data collection

Nearly always, a floating drilling operation is a wildcatting venture in an area previously undrilled. It represents money spent with the intention of justifying additional expenditures in the development of the area. With this purpose established, the floating drilling operation becomes more important than just investigating the subterranean prospects for oil and gas. Its aims should be extended to obtaining detailed data on ocean-floor sediments, currents, waves, water temperatures and other oceanographic conditions critical to further operations. During drilling operations, the opportunity is there to collect all these other data at relatively low cost.

Data on winds, seas, waves, and vessel motion should be recorded on a daily basis. In addition, vessels should be equipped with meters to record periodically current and temperature profiles at each well site.

Future design and construction of platforms, pipeline, and other production facilities will benefit from these data compilations. Even additional floating drilling operations can be conducted more efficiently.

Minor expenditures for instrumentation can also provide useful information for the evaluation of equipment and systems. Strain gauges on marine-riser joints have been used to evaluate fatigue damage and to locate stress concentrations.[3 6] Counters mounted on pneumatic tensioning units can record cycles which, when used with the tensioning information, can be the basis for the evaluation of wire-rope types and the retirement of lines.

The design of all of the subsystems for operating in deeper water requires more careful selection and use of the basic oceanographic and equipment-performance data.

REFERENCES

1. Adams, R. B.: "Accuracy of the Taut-line Position Indicator for Offshore Drilling Vessels". Petroleum-Mechanical Engineering Conference, ASME, Paper No. 67-Pet-5, May 1967.

2. Childers, M. A., Hazlewood, G., and Ilfrey, W. T.: "Marine Riser Monitoring with the Acoustic Ball Joint Angle-Azimuth Indicator". Offshore Technology Conference, OTC-1386, April 1971.

3. Harris, L. M. and Ilfrey, W. T.: "Drilling in 1300 Feet of Water—Santa Barbara Channel, California". Offshore Technology Conference, OTC-1018, May 1969.

4. Hartdegen, F. W. III: "The Application of Underwater TV and Video Tape Recording in Supervising and Documenting Offshore Operations". Offshore Technology Conference, OTC-1176, April 1970.

5. Bennet, P. B.: "Simulated Oxygen-Helium Saturation Diving to 1500 feet and the Helium Barrier". Offshore Technology Conference, OTC-1435, April 1971.

6. Tidwell, D. R. and Ilfrey, W. T.: "Developments in Marine Drilling Riser Technology". ASME Petroleum Mechanical Engineers Conference, 69-Pet-14, Sept. 1969.

16

Drill-Stem Testing

Drill-stem testing has been used by the oil industry for more than 30 years as a means to evaluate the productive potential of a well prior to final completion. Its original purpose was primarily to test in the open hole and to determine if an investment in the casing and the completion of the well were justified. The tests were flowed through the drill pipe used in drilling. Hence, the term drill-stem test arose.

Through the years, the cost of test tools and rig time increased at a more rapid rate than the cost of casing. The amount of drill-stem testing has tended to decrease and to be replaced by production testing. The development of the through-tubing perforating tools has also offered great advantages in production testing.

The interest in drill-stem testing has renewed, however, with the advent of mobile offshore drilling rigs. Now, the economic incentive is even more attractive. These tests often evaluate the need for a $5 to $25 million development-well platform.

Data available from drill-stem tests

Pressure recorders installed below the test-tool valve are the primary source of data for reservoir evaluation. They provide directly: (1) the initial formation pressure; (2) the wellbore flowing pressure at the tested rates of production; (3) the pressure build-up information after the test has been completed; and (4) the hydrostatic pressure of the drilling fluid. If the well is flowed to the surface, details are available on producing rates and fluid content. Most test tools are also provided with an internal sampler that collects fluid at bottom-hole conditions.

General rules for safety

Drill-stem tests can be conducted safely from floating drilling vessels, but compensation for vessel motion and consideration for oceanographic conditions are needed to assure full safety. The tests should be limited to cased holes and, preferably, with tubing used as the test string.

Certain restrictions on the timing of the test operations are important. In general, some operations should be conducted only in the daylight hours. These include: (1) opening the test tool; (2) flowing the well to the surface; (3) reversing the fluids from the test string; and (4) unseating the packer and pulling the tools. The subsea blowout-preventer stack should be equipped with at least one set of rams of the same size as the test string. And, the annulus area should be kept closed except when the tools are being manipulated.

Just as in all other floating drilling operations, provision must be made for alternate methods of performing certain steps in the test procedure. Alternate methods are needed for closing in the well and for reverse circulating the fluids from the test string.

Surface pressures should be kept to a minimum and full precautions taken for pollution and fire control. Alertness is required of all personnel involved in the design and the operation of the test. The supervisor in charge of the test operations should be prepared to discontinue the test immediately if:

1. Surface pressure becomes excessive. During and immediately following the unloading of the water cushion and the rathole mud, the surface pressure is determined by the amount of fluid to be unloaded. If the pressure cannot be reduced below 1,000 psi after production starts, the test should be stopped. Preferably, surface pressure should be reduced to about 500 psi.

2. The surface facilities become overloaded by rate or storage capacity. This applies to both liquid and gas facilities.

3. The annulus-pressure behavior is erratic. Some variation in pressure is normal from the effect of temperature and tubing pressure.

4. At any time the supervisor is concerned about safety because of weather, vessel motion, doubtful integrity of any equipment, abnormal presence of hydrocarbon vapors or hydrogen sulfide, or *any* reason he deems unsafe for the personnel or operation.

Types of equipment available

The tester valves of the two major service companies have both been used extensively on floating drilling vessels. Both have proven reliable. The Halliburton Hydrospring tester is actuated by simple up-and-down motion. In the up position, the valve is closed; in the down position, it is open. It has a hydraulic time delay built in, to prevent premature opening of the tool during packer-setting operations or if an obstruction is hit while running the tool.

The Johnston MFE tester is also actuated by up-and-down motion but requires a cycling motion to open and to close.

When the tool is first set down on the packer, it assumes the open position. A pick-up and set-down motion cycles the tool into a closed position. It can be alternately opened and closed by this same cycle. It too has a built-in time delay on opening.

Three basic types of reverse-circulating valves are currently in use—the rotating tool, the impact sub, and the pump-out plug.

The rotating reverse-circulating tool is designed so that rotation opens ports in the test string to permit circulation. Some designs incorporate a valve that closes the test string below the ports prior to opening the test string to the annulus. This is an added safety feature that should be utilized in all tests. Experience has shown the rotating reverse-circulating tool to be reliable, and it is the preferred of the three types.

The impact sub is simply a sub with a side port blocked by a sealing plug. It is opened by dropping a bar from the surface to shear off the sealing plug. The tool is generally reliable but can be difficult to operate if heavy mud, emulsion, or low-gravity oil above the tool cushions the dropping bar. Then, too, no equipment can be installed above the impact sub that would block the passage of the dropping bar.

The pump-out plug is merely a shear disc that is sheared by pressure applied to the test string. With high annulus hydrostatic heads imposed by heavy muds and with the probable presence of oil in the test string, high surface pressures can be required to shear out the plug.

Because of the possible malfunction of any one tool and because of the pollution and safety aspects of pulling a test string that has not been reversed, two types of reversing tools should be used in all drill-stem tests.

A test-string safety valve has been designed specifically for testing from floating drilling rigs. Current designs are actuated by vertical motion of the test string. When the test string is lowered, the valve closes. The tool serves two purposes. If during a test, a test string separates near the top, the string will drop and the weight of the string will close the safety valve. Or, in an emergency, the test string may be lowered purposely to close the valve.

A test-string safety valve should be used in all drill-stem tests from floating drilling vessels.

The subsea test tree (SSTT) is designed to land in the subsea wellhead or the blowout-preventer stack and to provide a means of closing in the well at the ocean floor. It is equipped with dual, hydraulically operated, fail-safe valves, with hang-off shoulders, and with a disconnect feature.

SSTT valves are designed for hydraulic opening from on board the drill vessel. If hydraulic control is lost, the two valves close. This feature is desirable insurance against a sudden, unprepared departure of the floating vessel.

The hang-off shoulders have been provided to support the test string on the ram-type blowout preventers, or in the wellhead, during testing. With one of the preventers closed and sealed around the SSTT, annulus pressure can be read on the surface through either the kill or choke line. A positive pressure of 500 to 1,000 psi provides accurate monitoring of the annulus.

The disconnect feature of the SSTT provides a safe convenient way of disconnecting the drill vessel from the subsea assembly even though the drill-stem-test tools are set and in place.

As mentioned, the SSTT may be landed in the wellhead or on the ram-type preventers. If the drill-stem-test assembly

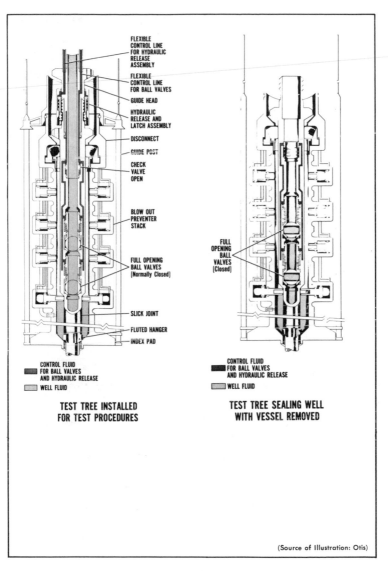

FLEXIBLE CONTROL LINE FOR HYDRAULIC RELEASE ASSEMBLY

FLEXIBLE CONTROL LINE FOR BALL VALVES

GUIDE HEAD

HYDRAULIC RELEASE AND LATCH ASSEMBLY

DISCONNECT

GUIDE POST

CHECK VALVE OPEN

BLOW OUT PREVENTER STACK

FULL OPENING BALL VALVES (Normally Closed)

SLICK JOINT

FLUTED HANGER

INDEX PAD

FULL OPENING BALL VALVES (Closed)

▇ CONTROL FLUID FOR BALL VALVES AND HYDRAULIC RELEASE

▨ WELL FLUID

TEST TREE INSTALLED FOR TEST PROCEDURES

▇ CONTROL FLUID FOR BALL VALVES AND HYDRAULIC RELEASE

▨ WELL FLUID

TEST TREE SEALING WELL WITH VESSEL REMOVED

(Source of Illustration: Otis)

FIG. 16.1 Sub Sea Test Tree

167

is designed for testing with the SSTT landed on the wellhead, then the SSTT is in its lowest position because the landing base will not pass through the wellhead. It is better to design the system with the SSTT to land on a ram-type preventer for the testing position. When it is so designed, the rams can be opened and the SSTT landing base lowered down to the wellhead. In this latter case, the system can be so spaced out that the safety valve closes with this additional downward movement. This permits purposely closing the safety valve. It also provides additional safety by having an additional means of closing off flow from the bottom.

Planning the drill-stem test

As in most other floating-drilling operations, good planning for the drill-stem test improves efficiency and safety. The geologists and engineers who will analyze the results must know the full range of potential data and the limits of the data that can be collected with a single test. Advance decisions are needed on the most important data to be obtained, and the test should be planned accordingly.

All available information on the test interval should be collected and analyzed. Of first importance are the potential fluid content of the reservoir and the probable reservoir pressure. Basically, the flow rate should be controlled by the bottom-hole choke, and the type of fluid to be produced determines the bottom-hole-choke size. Surface pressures should be limited as set out under "General Rules for Safety". For a gas-well test, this means that all pressure drop must be taken through the bottom-hole choke. The choke size is determined by the capability of the surface facilities to handle the flow rate. Commonly, this is only about 2.5 to 3.5 MMcfd and requires choke sizes as small as ⅛-in.

More control can be exercised at the surface on oil-producing reservoirs and still stay within surface-pressure limitations. Until the zone has been tested one time and until good information is available on the producing rates that can be safely handled aboard the vessel, it is probably best to limit producing rate to 1,000 to 2,000 b/d. Generally, a ¼-in. bottom choke is a good starting place for an oil test.

When using a ¼-in. or smaller choke, the length of water cushion is unimportant for protection of the formation. The larger the water cushion, the larger the surface pressures required to unload it, particularly so in the case of gas tests. For this reason water cushions should be limited to about 1,000 ft. In deep wells and with high mud weights, larger water cushions may be required to protect the test string from collapse. If this is the case, care must be exercised throughout the test to protect the test string from collapse. The problem can be quite acute for limited-capability, high-pressure gas zones.

Either slotted or perforated tail pipe is satisfactory for screening the choke. Hole sizes or slots in the pipe should be no larger than the bottom-hole-choke size.

Heavy-wall tubing with API threads is ideal as a test string when working from a floating vessel. The extra wall thickness is recommended for additional safety against burst or collapse. The additional strength in collapse permits the use of smaller water cushions, hence, lower surface pressure in unloading the water cushion. The API threads are recommended for additional protection against thread leaks after repeated usage.

The drill-stem-test string should be reserved only for testing. It should never be used for drilling operations. Frequent inspection and replacement are essential for reliability.

Example test-string design and procedure

Tables 12 and 13 illustrate a typical drill-stem-test assembly and procedure. The tools selected for the test determine space out and manipulations, hence, the assembly and procedure. The particular tools used in this example are furnished by Halliburton Services; tools of other companies have also proven reliable and satisfactory.

Example alternate test-string design and procedure

Tables 14 and 15 show an alternate drill-stem-test assembly and procedure using tools available from Johnston. A variation in the method of landing the subsea test tree is also included. With the Johnston equipment, the subsea tree may be landed on the wellhead or on the pipe rams, as desired.

Surface equipment

Steel hoses are used to connect the well from the control head on top of the test string to the choke manifold on the rig floor. The choke manifold is rigged with dual flow paths to permit changing chokes without shutting-in the well. With the exception of controlling the unloading of water cushion and rathole mud, regulation of the flow should be through the use of bottom-hole chokes to minimize surface pressures. On oil-production tests, however, some restriction can be taken at the surface to vary producing rates without exceeding safe surface pressures.

Manifolding which permits flow from the chokes to either the separator, the mud pit, or overboard simplifies unloading operations. The water cushion can be unloaded overboard, the rathole mud returned to the mud system and hydrocarbons directed to the separator. Some storage-tank space is essential for initial mud, water and oil flows and for gauging stabilized flow rates. The use of a burner can reduce the requirements for surface storage and can be helpful in conducting a test with minimum risks. Metered volumes on produced oil are not reliable enough to permit exclusive use of the burner, however, and after the well has stabilized, tank gauge readings are needed. Meter readings thus calibrated against the tank gauges can be useful in estimating the overall test production.

Two 100-bbl tanks (or one 200-bbl tank with two compartments) have proven adequate when operating with the burner. If a burner is not used, at least two 250-bbl tanks are needed and total production from the test is limited to this storage capacity.

Flare lines should be provided at both the bow and stern of the vessel to permit flaring in either direction. Overhead stack lines in the mast should never be used because of the danger of separator carryover and the consequential fire hazards.

Certain analysis equipment should be provided on board to obtain maximum data from the test. Sample bottles should be available for frequent fluid sampling; centrifuges are needed for shake-outs; and devices for heating and treating the emulsions are helpful, particularly, with high pour-point crudes. Hydrometers are normally used for API-gravity determinations.

Table 12

Example Drill-Stem-Test Assembly

OPERATOR: ABC OIL & GAS COMPANY
BASIC DATA:
    WELL NAME: ABC NO. 9-3-2
      WELL LOCATION: Off Utopia
DRILLING RIG: ZOOTCO VIII
    WATER DEPTH: 1050'
DST NO. 1 TEST INTERVAL 8020-8040'
    TOTAL DEPTH 10,000
CASING PROGRAM: 7-inch, 32 lb/ft set at 10,000
TEST STRING: 3½'' EUE, 12.9 #/ft N-80,
    DST PACKER DEPTH 7980'
MUD WT: 9.8 #/gal GRADIENT 0.509 psi/ft
HYDROSTATIC PRESSURE AT PACKER: 4062 psi
WATER CUSHION: 1000'
    ANTICIPATED FORMATION FLUID: Oil

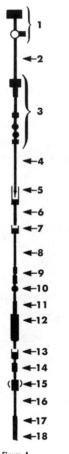

Figure A

TEST STRING ASSEMBLY (top to bottom): (Fig. A)

1. Swivel Control Head

2. Test String

3. Subsea Test Tree (Landing Collar Above Valves)

4. Test String

5. Safety Valve, 5-foot stroke, Final Position: Open

6. Test String, 180 feet

7. Slip Joint, 5-foot stroke, Final Position: Partially
   Closed

8. Drill Collars, 6 4¾-inch OD

9. Pump-out Reversing Valve, $\Delta P = 2000$ psi

10. Shut-in Rotating Reversing Valve

11. Pressure Recorder, 4500 psi, 48-hour clock

12. Hydrospring Tester, ¼-inch bottom hole choke

13. Slip Joint, 10-foot stroke, Final Position: Closed

14. Safety Joint

15. Drill Stem Test Packer

16. Perforated Tail Pipe

17. Pressure Recorder, 4500 psi, 48-hour clock (Inside)

18. Pressure Recorder, 4500 psi, 48-hour clock (Outside)

CALCULATION OF SPACE OUT TO LAND ON PIPE RAMS: (Fig. B)

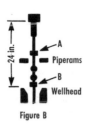

Figure B

Pick up to position Point A on pipe rams	14 feet
Collapse of packer (Item No. 15)	1 foot
Collapse of bottom slip joints (Item No. 13)	10 feet
Collapse of Hydrospring (Item No. 12)	2 feet
Collapse of upper slip joint (Item No. 7)	4 feet
Collapse of safety valve (Item No. 5)	0 feet
TOTAL LOSS IN DRILL STRING IN OPERATING POSITION	31 feet
REQUIRED FREE LENGTH OF DRILL STRING (7980' + 31')	8011'

Table 13
Example Drill-Stem-Test Procedure

1. Calibrate all bottom-hole pressure gauges prior to installation in test string.
2. Run assembly below SSTT, as shown in Table 12 items 4-18. (After 1,000-ft run, install water cushion).
3. Install SSTT. Test operation of valves.
4. Run upper assembly (above SSTT), except swivel control head.
5. Land SSTT in wellhead, verify measurements.
6. Install swivel control head.
7. Pick up 15 ft, close pipe rams, lower assembly 1 ft to land Point A on pipe rams.
8. Mark pipe in reference to rig floor. Observe vessel motion.
9. Open pipe rams and pick up and mark 17 ft below first mark. (Amount of tool collapse).
10. With the lower mark in exact position, set packer.

11. Lower assembly until upper mark is 1 ft above position. Close rams and lower SSTT to land Point A on pipe rams.
12. In 3-4 minutes tool should open. Some surface indication of a slight jar in the pipe or a blow on the test string may be observed when the tool opens. Mark the tubing "Point A".
13. Permit the well to flow 3-4 minutes to release the hydrostatic pressure below the tester valve.
14. Open the pipe rams, raise the test string 12 ft, close rams and land Point B on pipe rams. The tester valve should be closed in this position and indications of flow should diminish. Mark the tubing "Point B".
15. Leave tool closed one hour to obtain initial shut-in pressure. During this time connect the steel hoses between the swivel control head and the floor manifold.
16. Open the pipe rams, lower the test string to land Point A on the pipe rams. Close rams and land on Point A. Pressure the annulus to 1,000 psi. Observe behavior of annulus pressure throughout test.
17. The tester valve should open in 3-4 minutes after the test string was lowered. Some surface indication of a slight jar in the pipe and/or a blow on the test string may be observed when the tool opens.
18. Flow the well for desired test period.
19. To close the tool, bleed the annulus pressure, open the pipe rams, pick up to Point B, close rams and land on Point B. The tester valve should be closed in this position and indication of flow should diminish.
20. Open the pipe rams and without lowering the tubing rotate to the right 10 turns to open the rotating reverse circulating valve. (If the tool fails to operate, the pump-out reversing valve may be opened by applying pressure on the tubing string).
21. Reverse circulate well fluids from test string. Leave tester valve closed a minimum of 12 hours.
22. Open pipe rams if closed, pick-up test string to unseat packer, pull drill-stem-test assembly. Do this in daylight hours only.
23. As soon as test assembly is out of the hole, run bit on drill pipe to bottom. Circulate and condition hole.

Table 14
Example Alternate Drill-Stem-Test Assembly

OPERATOR: XYZ OIL COMPANY

BASIC DATA:
WELL NAME: XYZ NO. 3-1-6
WELL LOCATION: Off Shangri-la
DRILLING RIG: ZOOTCO IX WATER DEPTH 860'

DST 3 TEST INTERVAL 9521-39'
TOTAL DEPTH 9090'
CASING PROGRAM: 7-inch, 29 lb/ft set at TD
TEST STRING: 3½' EUE, 12.9 # /ft N-80
DST PACKER: 9480'
MUD WT: 9.8 # /gal GRADIENT: 0.509 psi/ft
HYDROSTATIC PRESSURE AT PACKER: 4830 psi
WATER CUSHION: 1000'
ANTICIPATED FORMATION FLUID: Oil

TEST STRING ASSEMBLY (top to bottom) (Fig. A)

1. Swivel Control Head
2. Test String
3. Subsea Test Tree (Landing Collar Above Valves)
4. Test String
5. Safety Valve, 5-foot stroke, Final Position: Open
6. Test String, 180 feet
7. Slip Joint, 5-foot stroke, Final Position: Partially Closed
8. Drill Collars, 6 4¾-inch OD
9. Pump-out Reversing Valve, ΔP = 2000 psi
10. Shut-in Rotating Reversing Valve
11. Pressure Recorder, 6000 psi, 48-hour clock
12. MFE Tester, ¼-inch bottom hole choke
13. Slip Joint, 10-foot stroke, Final Position: Closed
14. Safety Joint

Figure A

15. Drill Stem Test Packer

16. Perforated Tail Pipe

17. Pressure Recorder, 6000 psi, 48-hour clock (Inside)

18. Pressure Recorder, 6000 psi, 48-hour clock (Outside)

CALCULATION OF SPACE OUT TO LAND ON WELL-HEAD: (Fig. B)

Collapse of packer (Item No. 15)	1 foot
Collapse of bottom slip joints (Item No. 13)	10 feet
Collapse of MFE Tester (Item No. 12)	2 feet
Collapse of upper slip joint (Item No. 7)	4 feet
Collapse of safety valve (Item No. 5)	0 feet
TOTAL LOSS IN DRILL STRING IN OPERATING POSITION	17 feet
REQUIRED FREE LENGTH OF DRILL STRING (9480' + 17'):	9497'

Wellhead

SSTT IN
Test Position

Figure B

TABLE 15

Example Alternate Drill-Stem-Test Procedure

1. Calibrate all bottom-hole-pressure gauges prior to installation in test string.

2. Run assembly below SSTT, as shown in Table 14, items 4-18, (After 1,000 ft. run, install water cushion).

3. Install SSTT. Test operation of valves.

4. Run upper assembly (above SSTT), except swivel control head.

5. Land SSTT in wellhead, verify measurements.

6. Install swivel control head.

7. Mark pipe in reference to rig floor. Observe vessel motion.

8. Pick up and mark pipe 17 ft below first mark. (Amount of tool collapse).

9. With the lower mark in exact position, set packer.

10. Lower assembly until landing SSTT in the wellhead.

11. In 3-4 minutes tool should open. Some surface indication of a slight jar in the pipe or a blow on the test string may be observed when the tool opens.

12. Permit the well to flow 3-4 minutes to release the hydrostatic pressure below the tester valve.

13. Raise the test string 11 ft to close the MFE tool, then set the SSTT back down on the wellhead. With the tester valve closed, indications of flow should diminish.

14. Leave the tool closed one hour to obtain initial shut-in pressure. During this time connect the steel hoses between the swivel control head and the floor manifold.

15. Raise and lower the test string 11 ft again to manipulate the MFE tester tool. Close the pipe rams and pressure the annulus to 1,000 psi. Observe behavior of annulus pressure throughout test.

16. The tester valve should open in 3-4 minutes after the test string was manipulated. Some surface indication of a slight jar in the pipe and/or a blow on the surface may be observed when the tool opens.

17. Flow the well the desired test period.

18. To close the well, bleed the annulus pressure, open the pipe rams, pick up the test string 11 ft, then set the SSTT back down on the wellhead. The tester valve should now be closed and indication of flow should diminish.

19. After indication that flow has ceased, pick up about 2 ft, rotate the test string 10 turns to open the rotating reverse-circulating valve. (If the tool fails to operate, the pump-out reversing valve may be opened by applying pressure on the tubing string.) Lower the SSTT back down on the wellhead.

20. Reverse circulate well fluids from the test string. Leave the tester valve closed a minimum of 12 hours.

21. Open the pipe rams (if closed), pick up the test string to unseat the packer, pull drill-stem-test assembly. Do this in daylight hours only.

22. As soon as the test assembly is out of the hole, run bit on drill pipe to bottom. Circulate and condition hole.

17

Control
and Removal of
Oil Spills

Offshore petroleum exploration and production operations can be safely conducted without danger to human life, can be compatible with other marine activities, and can preserve the natural beauty and environment of the ocean and beach areas. Equipment and practices designed for safety and reliability are the first line of defense against oil spills and pollution. Should a spill occur, however, advance planning can reduce its severity and minimize clean-up time, damage, and costs.

As does most planning, oil-spill-control planning must start with a review of methods, materials, and manpower available. Most authorities advocate this course of action:

1. Control the source of the spill
2. Contain the spill
3. Pick up of the major part of the oil
4. Disperse remaining oil

Each spill presents individual circumstances, and contingency plans must provide flexibility. Spills may occur on the high seas remote from any land mass or in the immediate vicinity of heavily populated beach areas. The volume and type of oil and sea conditions also affect equipment and methods that should be employed.

Every contingency plan should consider the location of equipment, supplies and manpower, the involvement of concerned governmental agencies and press representatives, and the advance evaluation of methods and materials.

Containment methods and equipment.

Innumerable barriers, or booms, have been offered by various manufacturers. Most have only limited application, and are ineffective in rough seas or with high currents. Most can be effective in calm water and can protect harbor estuaries, stream outlets and other confined areas.

Effective barriers must be constructed of materials resistant to oil degradation and capable of withstanding the environmental conditions of sun, temperature, and sea water. The materials must resist tear or abrasive damage, and also be capable of withstanding long periods of storage.

The barrier must ride with the sea, withstanding wind, current and waves while moored in position or under tow. It must have sufficient height both above and below water to prevent overflow and underflow of the oil. It should be of minimum bulk and weight to facilitate storage, transportation and installation.

One design that appears to offer promise is a barrier developed by Ocean Systems, Inc., under contract to the U. S. Coast Guard.[1] The system is designed to contain oil in 4- to 5-ft seas in combination with 20-mph winds and 0.7- to 1.0-knot currents.

The barrier design is based on the use of flexible polyurethane foam, with a "dynamic keel" that imparts high static and dynamic stability. The barrier consists both of a non-water-absorbing foam package that provides the buoyancy and a surface barrier. A water-absorbing foam package provides a submerged barrier and serves as a "dynamic keel". The two packages are connected into an integral unit that can be compressed to approximately 20% of the original volume for storage and transportation. The resiliency of the foam material causes the barrier to resume its original shape and size after packaging restraints are released. No compressors, pumps or other mechanical support equipment are required.

The barrier is of modular construction. Each module is 17 ft long and units can be assembled in any desired length. The upper portion extends 2 ft above the water level, and the lower 4 ft below. Seals prevent seepage between modules, and connectors keep the modules together. A dacron strength belt, located in the center of pressure of the bottom section, maintains the barrier's structural integrity under high seas. The strength belt's ultimate strength exceeds 200,000 lb.

Barriers are used in one or more of three ways. They may be used: (1) to surround and contain a spill; (2) as a sweep and towed to concentrate the spill; or (3) to protect an area (such as a harbor) from invading oil. When moored, they can be attached to anchored buoys, to holding vessels or to the shore. When towed, two vessels usually draw the barrier in a parabolic configuration.

Experimental work has also been conducted by Texas A & M University on pneumatic barriers for containing oil spills.[2] The bubble-generated current was found an effective means of containing oil on water. However, under currents of two knots or breaking wave conditions, the large quantities of air required probably make the system uneconomic for most applications.

Skimmers, oil-removal equipment

Once the oil spill is contained or concentrated with barriers, it must then be picked up and transported to a disposal site. The oil can be gathered by suction nozzles, weir skimmers, sorbents, or several other special devices designed for this purpose.

Suction nozzles work efficiently when used on thick layers of oil, but their operation is upset by waves.[3] When used on thin or patch oil slicks, they collect a disproportionately large amount of water, so that huge storage and separation facilities become necessary. Weir skimmers are intended to allow the surface layer of oil to flow by gravity into a container from which it can be removed. They work admirably on a thick layer of oil, provided the waves and current are small. However, like suction nozzles, they also collect water in large quantities if the oil layer is thin or the waves are large.

Sorbents are difficult to distribute and to recover when used over a large area. Then, too, in the case of a large spill, large amounts are required and the handling and disposal of the used material can be expensive. Gelling agents present most of the same problems as sorbents.

Other systems and concepts for oil recovery either under study or in development include: a free vortex induced by an impeller to concentrate the oil for pump pick-up; a rotating disk preferentially wet by oil and collected by wipers; and numerous other mechanical devices. Some appear to have merit.

Chemical dispersants and application equipment

Chemical dispersants are the best method for removing: (1) small spills; (2) widely spread thin layers of oil; (3) oil on rough seas; and (4) oil spills immediately threatening damage to property or life. Although dispersants are restricted or prohibited in many locations by regulatory agencies, they still offer the safest and quickest way to immobilize the danger of damage from the spill.

Chemical dispersants now in use work toward removing the oil from the water surface by formating an oil-in-water dispersion. Because the oil does remain in the sea, the use of dispersants is not as desirable as removing the oil completely. As mentioned, however, dispersants have application where other methods are inadequate and are highly preferable to some consequences that might occur otherwise.

Canevari[4] has very simply explained the working of the chemical dispersants. When a surface-active agent (surfactant) is applied to an oil film on water, it lowers the interfacial tension because of its amphiphatic nature; i.e., partly oil soluble (lipophilic) and partly water soluble (hydrophilic). By reducing interfacial tension in this manner, it enhances the generation of interfacial area upon the application of mixing energy. A more subtle requirement of the surface-active agent is prevention of coalescence of the droplets once they are formed. In essence, the surfactant acts to fend and physically parry droplet collisions. This same property also reduces the tendency of droplets to stick to and thereby wet an immersed solid surface. The vital component of any oil spill dispersant is, therefore, the surface-active agent.

The dispersant then acts solely as an agent to enhance the formation of oil droplets. It does not "weight" the droplets in order to sink them. It does not solubilize the oil into the water. It simply promotes oil-in-water dispersion.

Canevari offers the following benefits for chemically dispersing and removing oil from the surface of the water:

1. The rate of biodegradation of the oil is increased.

2. Damage to marine fowl is avoided since oil is removed from the water surface.

3. The fire hazard from the spilled oil is reduced by dispersion of the oil several feet into the water column.

4. The spilled oil is prevented from wetting solid surfaces such as beach sand, shore property, etc.

5. The formation of a tar-like residue from an oil spill is prevented.

A number of dispersant chemicals are currently offered for oil-spill cleanup. Good, independent evaluation is needed to compare their relative merits. Testing should evaluate: (1) the dispersing qualities of the product with various grades and types of oil over a range of sea-water salinities and temperatures; (2) the toxicity level on various species of marine life; and (3) biodegrability, that is, the susceptibility of the chemical to oxidation by bacteria which live in the sea.

Information available indicates that Enjay Chemical Co.'s Corexit 7664 is an acceptable dispersant for this use. Others on the market are probably as good. Enjay recommends "a use level of 2 to 10 parts Corexit 7664 per 100 parts of oil depending on the film thickness", with its application in a 4 to 6% solution in water.

Dispersants can be applied by several techniques. In the final analysis, the chemical should be spread in a fine uniform spray and subjected to maximum agitation. Aircraft of the crop-spraying type do an effective distribution job, particularly where rough seas provide the agitation. In calm water, boats with spray equipment are more effective because of the agitation supplied by the wake of the boat.

Fixed and portable fire-fighting systems can be effective on drill vessels, workboats, and even at dockside for distributing the chemical and providing the required agitation. The units should include chemical-injector pumps for blending the dispersant into the main water stream. As the mixture is applied to the spill area, the energy developed by the pumps provides agitation; later, by cutting back the chemical additives, additional passes can be made with the fire hoses simply for additional agitation.

The small hand-operated pump units and pressurized units (1- to 5-gal size) are a good means of distributing chemical to minor spills. They are also useful in cleanup operations around a harbor area where access to the spill is difficult for larger units. The hand units can be particularly useful around the rig to negate small oil accumulations in the moon-pool area or in sumps.

Other chemicals, methods of oil removal

The burning of oil spills offers an attractive means of eliminating them. However, the opportunities of safely using this technique are rare. First, the accumulation of oil must be isolated from any property, shoreline or ships that might be damaged. The accumulation must be of sufficient size and continuous enough to support combustion. Commercial burning agents assist in the ignition and in sustaining combustion for the heavier crudes and fuel oils. However, these can be expensive and relatively ineffective unless conditions are favorable.

Chemical sinking agents are sometimes used to create a high-density compound or agglomerate to remove the oil from the water surface. Generally, these agents postpone the problem rather than cure it and are not permitted in some areas.

Milz and Fraser[5] report a surface-active chemical agent which possesses a spreading force greater than that of oil. Hence, it prevents the spreading of oil on water. And, in most cases, it causes an existing oil slick to contract to much smaller area. The advantage of this agent is to facilitate the recovery of the oil by sorbents or other pickup methods.

Beach-cleanup methods

The failure or inability to control and recover an oil spill at sea invariably results in the oil coming ashore. Sometimes, this has serious economic and ecological consequences. Like most of the cleanup operations at sea, the cleanup on the shore line is best attacked with an integrated approach. Still, it usually ends requiring a large amount of hand labor before the beach is fully restored.

Oil absorbents, such as straw, sawdust, peatmoss, kelp, compost, polyurethane foams, etc., appear to be practical and effective in the shoreline operations. Union Oil in combating the historic Santa Barbara spill in 1969 used straw blowers to apply straw directly to the beaches to absorb the oil as it washed ashore. Sorbents can be also distributed over the spill area near the shore line. These contact the oil prior to reaching the beach sands but, in most cases, are more expensive to apply offshore. Cleanup operations eventually involve the pickup of the oil-soaked sorbents and disposal by burning or transporting away.

Sartor and Foget[6] have reported research studies and actual beach-restoration operations using selected earthmoving equipment to handle various types of oil-contaminated beach sands. Their evaluation included motorized graders and scrapers for collecting and removing the contaminated sands and sorbents and the equipment for loading and screening systems. This study offers good information on techniques, costs, effectiveness, and equipment modifications needed for efficient removal. In general their study concluded:

1. A motorized grader and motorized elevating scraper

working in combination provide the most rapid means of beach-restoration when oil penetration is limited to less than 1 in. For greater oil penetration, the motorized elevating scraper operating singly is more efficient.

2. The oil-removal effectiveness was greater than 98% for all restoration procedures.

3. A front-end loader mounted on a crawler tractor was the most inefficient apparatus tested. In addition, more spillage occurred with its use than with any other equipment.

4. A non-elevating motorized scraper will not operate efficiently on beach areas unless a tracked prime mover is used as the principal source of power or as a pusher to assist in loading.

5. Beach-restoration operations on backshore areas are more difficult due to the looseness of the sand.

6. Conveyor-screening systems can be effectively utilized to: (a) load oil-contaminated material into trucks; (b) separate oil-sand pellets from clean sand; and (c) partially separate oil-contaminated debris (straw, kelp, seaweed) from sand.

7. The mixing action that occurred in the cutting and pickup of a thin film of fresh oil and the underlying clean sand results in a uniform oil-sand mixture. It is not possible by screening to separate oil-contaminated sand from clean sand.

8. Costs for removing a thin layer of oil from the beach tidal zone ranged from $108/acre to $1,540/acre for restoring the area, depending on the type of equipment used.

Hot-water and steam apparatus seems to be the only practical, effective method of cleaning rocks, pilings, docks, and other hard surfaces that have been oil-coated. Boats, machinery and other valuable private property are best cleaned by the owners and costs settled individually.

Public, governmental relations

Damage to the industry image can be more costly than actual physical damage resulting from an oil spill. Thorough preplanning is warranted for maintaining good public and governmental relations.

The use of chemicals (such as dispersants) particularly requires advance attention. Governmental agencies (fish and wildlife commissions) and schools of the marine sciences should be included in joint evaluations of the effect of various chemicals on marine life. Results of these evaluations should dictate the selection of chemicals in the clean-up operation. This information should be made readily available to the press.

Contingency plans should be developed with the cooperation of concerned governmental agencies. The plans should provide for the notification and enlistment of aid from these agencies in the event of an oil spill. Provision should be made to notify the press and to assign personnel to work with the press and see that full, correct information is readily available. An informed press is the surest way to an informed public. An informed public is generally appreciative of man's problems and is quick to recognize sincere effort and is surprisingly tolerant of complications and damages if sincere effort has been made.

REFERENCES

1. March, Frank: " 'Dynamic Keel' Oil Containment System". Joint Conference on Prevention and Control of Oil Spills, API-EPA-USCG, Washington, D. C., June 1971.

2. Basco, David R.: "Pneumatic Barriers for Oil Containment Under Wind, Wave, and Current Conditions". Joint Conference

on Prevention and Control of Oil Spills, API-EPA-USCG, Washington, D. C., June 1971.

3. Betts, W. E., Fuller, H. I., and Jagger, H.: "An Integrated Program for Oil Spill Cleanup". Joint Conference on Prevention and Control of Oil Spills, API-EPA-USCG, Washington, D. C., June 1971.

4. Canevari, Gerald P.: "Oil Spill Dispersants—Current Status and Future Outlook". Joint Conference on Prevention and Control of Oil Spills, API-EPA-USCG, Washington, D. C., June 1971.

5. Milz, E. A. and Fraser, J. P.: "A Surface-Active Chemical System for Control and Recovery of Oil from Ocean Environments." Offshore Technology Conference, OTC-1376, April 1971.

6. Sartor, James D., and Foget, Carl R.: "Evaluation of Selected Earthmoving Equipment for the Restoration of Oil-Contaminated Beaches". Joint Conference on Prevention and Control of Oil Spills, API-EPA-USCG, Washington, D. C., June 1971.

18

Operations for Maximum Safety and Efficiency

Technology and equipment are available today for safe, efficient floating drilling operations. Policies even in early planning stages should take advantage of past experience to insure design for full reliability in the systems and the operation. Policies based on full safety invariably include standards conducive to maximum efficiency and to the lowest total cost.

Basic management policies

Standards set forth in this text have been established through experience and have proven in general the safest, most economical approach. Using this information as a background, basic criteria can then be established:

1. Factors of safety in equipment design should be sufficient to retain full capability under design load conditions even after normal wear. They should be sufficient to safeguard against some inaccuracy in the selection of the design conditions.

2. Exacting standards for inspections and maintenance should be established to insure handling, servicing, repair and replacement that retains the original reliability of the equipment.

3. Redundancy should be provided and offer alternate methods of operation and control in an emergency and should not be utilized for the purpose of delaying repair to equipment.

4. Operating practices and emergency procedures should be adopted and established in writing (preferably in manual form) for the guidance of personnel directly associated with the operation.

Manual on operating practices and emergency procedures

An effective manual is one that is complete and accurate, yet convenient to use. It should be written in simple, easily understood language, carefully indexed, and frequently revised to maintain current information. It should be prefaced with a clear concise management statement of policy in regard to safety, well control and protection for the environment. Reference telephone numbers and addresses should be included for all key company offices and personnel, hospitals, government and regulatory agencies, and emergency and rescue groups.

The major sections of the manual should include standards on vessel security, on well security, on environmental control, on drill-stem testing, and on supply and support facilities. Each should include design considerations, operating practices, inspection and maintenance standards, and emergency procedures. The design considerations should set forth equipment specifications based on the established factors of safety and redundancy requirements. The operating practices should set the standards on the use of the equipment and materials and on the monitoring and control of the performance of both equipment and personnel. The inspection and maintenance standards should specify the frequency and type of inspections and standards for the care, repair and replacement of material and equipment. Emergency procedures should set the pattern for action in the event of an actual emergency condition.

Training and development of personnel

The most serious problem currently facing industry in floating-drilling operations is the shortage of adequately trained personnel. The shortage of rig crews has developed from two causes—the rapid expansion of floating-drilling operations and the reluctance of the industry to pay adequate wages. The argument has been offered in the past that unless labor costs were controlled the drilling contractor would rapidly price himself out of business. That argument may have had some validity in past land operations where crew costs were a major part of the daily total. With floating drilling operations commonly costing in excess of $25,000 per day, the crew costs are only a minor part of the total. Operators can well afford additional daily rental rates to provide adequate wages for top crew performance. More important, no operator can afford the cost of crew inefficiencies brought about by inadequate wages.

The shortage of personnel is not limited only to rig crews. Most operating companies are seriously short of technical and supervisory personnel. These companies have usually had only erratic involvement in floating drilling. Hence, they have been hesitant to develop personnel that might be difficult to place during periods of inactivity. Changing levels of activity are not new problems to the drilling industry. In the case of floating drilling, however, the incentives are more significant than in land operations. Working short-handed during periods of activity can be tremendously more expensive than carrying an extra load on the payroll during periods of inactivity.

Formal training schools, extra staffing and the extensive exposure of personnel to rig operations are the cheapest, fastest means of rapidly training needed personnel.

Daily operations

Efficient operations planning is particularly rewarding in floating drilling because of: (1) the sequential nature of the complex equipment handling in establishing subsea assemblies; and (2) the high daily costs. Drilling supervisors trained and experienced in planning ahead can employ their talents to maximum advantage and materially reduce well costs.

The continuing analysis of equipment-handling procedures offers frequent opportunities to reduce rig time and costs. One operator has published data showing the significant advantage of running the subsea blowout-preventer stack on an integral marine riser with the control system and choke and kill lines all installed simultaneously.[1] Table 16 sets forth the rig time required for running and pulling the subsea assembly as a function of water depth. For an average water depth of 910 ft, installation time for the integral riser has averaged 17½ hours with a range from 7 to 35 hours. For comparison, five wells drilled in an average water depth of only 455 ft and using separate kill and choke lines averaged about 50 hours installation time. Time to retrieve the integral riser system has averaged 14 hours, or only 65% of that for the non-integral riser system.

Numerous other opportunities to reduce rig time await the skilled supervisors.

TABLE 16
Published Blowout-Preventer And Riser-Handling Time
In The Santa Barbara Channel

Well No.	Drill Vessel	Water, depth, ft	Date	Running time, hours	Pulling time, hours
			Integral Riser		
OCSPO186 # 3	Wodeco IV	580	11-28-69	9.50	6.00
			12-09-69	7.00	6.75
			01-01-70	12.00	
OCSPO188 # 1	Bluewater II	941	12-05-68	25.00	14.50
			12-16-68	17.00	22.50
OCSPO188 # 2	Bluewater II	1,005	02-03-69	34.25	15.50
			03-28-69	27.00	15.50
			06-17-69	15.00	20.50
OCSPO190 # 1	Wodeco IV	1,299	09-25-68	32.25	16.75
			10-31-68	12.00	18.00
			12-25-68	14.00	10.00
			01-05-69	12.00	15.00
			01-14-69	14.50	21.00
OCSPO190 # 2	Wodeco IV	1,046	04-06-69	24.50	8.00
			05-30-69	9.00	9.75
			06-10-69	9.00	16.50
OCSPO190 # 3	Wodeco IV	1,245	08-01-69	14.00	8.00
			08-24-69	12.50	8.00
			10-16-69	14.00	7.50
OCSPO197 # 3	Bluewater II	730	07-24-69	25.25	25.75
			08-02-69	13.00	10.50
			09-02-69	21.25	8.50
			09-22-69	14.00	10.50
			10-07-69	15.00	15.50
OCSPO23 # 1	Wodeco IV	994	08-09-68	20.00	20.75
OCSPO235 # 2	Bluewater II	298	11-03-68	35.50	16.50
			11-14-68	15.00	11.00
Averages for			11-21-68	24.50	16.00
integral riser		910		17.50	13.85

			Non-Integral Riser		
OCSPO186 # 1	Bluewater II	629	03-06-68	68.00	18.00
OCSPO186 # 2	Wodeco II	279	08-15-68	34.50	17.50
			10-05-68	16.50	17.00
OCSPO197 # 1	Bluewater II	640	06-04-68	72.00	22.50
OCSPO197 # 2	Bluewater II	669	08-29-68	112.00	22.50
			09-07-68	18.00	28.50
OCSPO235 # 1	Wodeco II	310	05-04-68	85.50	18.75
			07-02-68	23.00	26.00
Averages for			08-01-68	16.00	21.50
non-integral riser		455		49.50	21.50

REFERENCES

1. Harris, L. M.: "Design for Reliability in Floating Drilling Operations". Second Offshore Technology Conference, OTC-1157, April 1970.

19

New and Future Developments in Floating Drilling Operations

The high daily cost of floating drilling operations offers an opportunity to develop new equipment and technology rarely present in drilling or any other industry. Large investments in research, development, and manufacturing can be justified for tools and equipment that improve reliability and reduce overall time on the well. With this incentive, technology and equipment are changing rapidly.

Surface-vessel motion-compensation units

A previous chapter discussed the inherent disadvantages of bumper subs in the drill string to compensate for vertical motion of the drill vessel. To reiterate, downhole motion compensation: (1) limits the flexibility of varying drilling weight on the bit without removing the drill string to change the amount of drill collars; (2) shortens the bit life in that complete compensation is not always achieved with bumper subs; (3) does not compensate for motion of the drill pipe, hence, creates problems of wear in the subsurface equipment, particularly, if blowout preventers are closed to contain surface pressure; (4) introduces weaknesses in the structural and

195

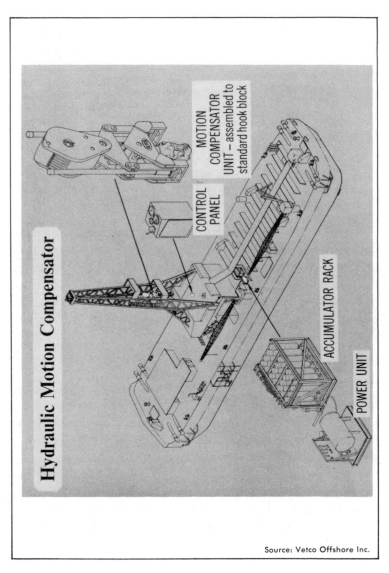

Hydraulic Motion Compensator

MOTION COMPENSATOR UNIT – assembled to standard hook block

CONTROL PANEL

ACCUMULATOR RACK

POWER UNIT

Source: Vetco Offshore Inc.

FIG. 19.1   Vetco Motion Compensator System

196

pressure integrity of the drilling assembly; and (5) adds to overall rig time and cost in the rental, repair and replacement of the bumper subs themselves.

One manufacturer has demonstrated the practicality of a surface motion compensation.[1] This is a pneumatic-hydraulic piston unit that mounts between the drilling hook and traveling block. It can be set by controls at the drillers position to maintain a set hook load. It strokes with the motion of the vessel to keep the weight on the bit constant. Variation in bit weight is possible by adjusting pressure on the hydraulic pistons. One unit currently in operation on the WODECO IV has been designed for a maximum hook load of 400,000 lb. and operates with a stroke up to 15 ft. Preliminary reports indicate the unit is performing satisfactorily. That is, it has maintained constant bit weight and has operated with a minimum of maintenance.

Units of this type offer the dual advantages of cost reduction through improved operating efficiency and improved overall safety and reliability.

Buoyant modules for marine risers

As water depth increases, the need for tensioning the marine-riser system increases. At least two rigs are currently equipped with pneumatic tensioning equipment capable of developing 360,000 lb of tension. These rigs are supposedly equipped to operate in water depths to 1,500 ft. With water depths of 2,500 to 3,000 ft being considered, the number of pneumatic tensioning units which will be required could easily double.

There are physical limitations to adding twice the number of pneumatic cylinders or larger units. Because of this, additional means of tensioning the riser or reducing the tension

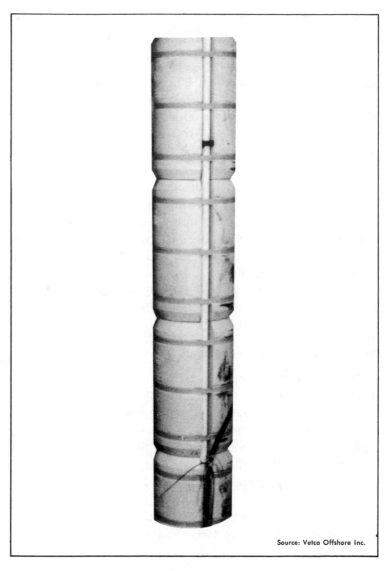

FIG. 19.2   Buoyant Modules Installed on a Joint of Integral Riser

198

requirements are under study. Some of the newer designs of riser connectors are considering weight to help minimize the requirements.

One of the more interesting new concepts is the addition of buoyancy to the marine-riser joints. Buoyancy units being tested are of a polyurethane-foam material, covered with fiber glass and fabricated in half cylinder form to attach around the marine-riser joints. These have been manufactured in 5-ft lengths and can be strapped in pairs on the riser pipe with stainless-steel bands using conventional steel-banding tools. Advantage of this approach is the convenience of applying the units in the field to existing marine-riser joints. Another advantage is the flexibility in the amount of buoyancy given by controlling the number of units installed per joint.

Tests have been made with the material in more than 1,000 ft of water in the Santa Barbara Channel. These particular units were designed to provide neutral buoyancy per joint, or to balance exactly the weight of each individual riser joint. The tests have been reported as satisfactory and the material has performed as expected.

Gas-diverter systems

Several operators of floating drilling rigs have reported concern about drilling in areas with over-pressured shallow gas sands. They wish to be able to close subsea blowout preventers on well flows before gas actually enters the riser pipe above the preventers. Their concern for this is justified. Once the gas gets into the riser pipe above the preventers, it is automatically channeled directly to the rig floor and, on most rigs, can be exhausted from the riser pipe by that means alone.

Equipment is available that can be installed on top of the marine riser and close around the drill string, diverting the gas away from the rig. Except in deep water (in excess of 700-800 ft), it is doubtful that human and equipment reaction times can be fast enough to do the job intended. In addition, the extra equipment on top of the riser increases installing and dismantling time for the marine-riser system. For these reasons, diverter systems may not be practical for most floating drilling rigs. They should be installed if the rig is working in an area of known trouble and if drilling is conducted with the diverter already closed in preparation for a flow.

Ram-position indicators

A previous chapter discussed the control system and the means available for monitoring the action of the rams of the subsea blowout preventers. The control lights on most hydraulic control systems are operated by pressure-actuated switches. Hence, they are a true measure of pressure being applied to the pilot valve. A blockage in the hose or a malfunction of the pilot valve could still create pressure on the pilot line. If so, it could give an erroneous indication on the control panel. As a cross reference, a counter on the control panel measures the volume of power fluid consumed. This is a further indication that the ram has shifted.

A limited amount of work has been done on the development of ram-position indicators to be mounted on the preventer. These would provide a positive signal to the surface as to the position of the ram. This additional means of monitoring the critical, remotely located well-control equipment should eventually become standard.

Re-entry devices

Devices designed for relocating the well bore are being used or considered for two different purposes: to assist in relocation of the well bore after the removal of the original drilling vessel; and to eliminate the currently used guideline system to guide equipment into the well bore prior to installation of the marine riser. In both cases the industry has only limited experience.

Old well bores can be located in several ways. Most common is installation of an acoustical transmitter (beacon) on the wellhead. Batteries of two years' life can be provided with the initial installation. If necessary, the beacons can be removed, serviced and reinstalled. Or, new beacons can be dropped periodically as the batteries need replacement. A better, less expensive method is to install on the wellhead acoustic reflector devices that only reflect signals. No batteries need be serviced.

Call-back marker buoys have been used extensively in the Gulf of Mexico. These devices can be activated acoustically to release a buoy and wire line that pop to the surface.

The acoustic systems used for vessel-position monitoring always have an acoustic beacon on the ocean floor which could be used for re-entry. Suppose a vessel has to disconnect and leave a location on account of weather. Or, suppose the vessel is suddenly, unexpectedly blown off location. The acoustic beacon for the vessel position-reference system is already on the ocean floor. It can guide the vessel back into the exact position.

A number of re-entry systems are currently being tested and developed for guiding the drilling systems into the well bore during routine drilling. Most are of the reflector type. They employ an acoustic transmitter on the bottom of the drill string, sending a signal to a preset reflector mounted near the well bore. The vessel is shifted as necessary to align the assembly with the well bore. It is doubtful that this system will ever be practical for anything other than short-term coring operations where blow-out preventers are not normally installed.

Explosive-set anchors

The U.S. Navy has under development an explosive-set anchor that ultimately should have application in floating drilling operations.[2] Objective of the Navy is an anchor for salvage operations that can be: (1) directly embeddable into sand, coral and mud seafloors without the necessity of dragging to embed and set it; (2) capable of developing a holding capacity of 160,000 lb horizontal force measured at the hawser of the salvage vessel; (3) operational and suitable for use in water depths 50 to 500 ft; and (4) practical for use aboard certain naval vessels in certain sea conditions.

The specifications designated by the Navy are not exactly those required by the petroleum industry. Still, the anchor could have extensive use with floating drilling vessels. The most interesting point of the Navy development is the tested holding power. Several have shown holding power in excess of 100,000 lb, even with a direct vertical pull. Anchors of this type set in a conventional spread mooring system should develop suitable holding power for drilling vessels.

The explosive anchor assembly consists of two major components: a reusable launch vehicle; and an anchor projectile. The overall assembly is 12 ft high, has a circular base 10 ft in diameter, and weighs about 12,500 lb. The launch vehicle supports and orients the anchor projectile prior to firing, then propels it into the seafloor. The anchor projectile embeds into the seafloor and becomes an anchor. It is not intended to be retrievable. A piston that inserts into the gun barrel forms part of the anchor projectile. Other essential features of the explosive anchor assembly include down-haul cables, a bridle cable, a mechanical cable-release device, and the ordinance system.

In operation, the anchor assembly is lowered to the seafloor and fired. As the anchor projectile ejects, it pulls the mechanical cable release freeing the beach-gear leg attachment from the side of the launch vehicle. The down-haul cables trail the anchor projectile into the sea floor. The launch vehicle then is retrieved for reuse in firing other anchor projectiles. The embedded anchor projectile is left in place with a cable to the surface.

Use of this system with a floating vessel would be similar to the use of anchor piles, except hopefully much less expensive. Explosive anchors with seafloor to surface anchor cables would be installed by a service boat in advance of the arrival of the drill vessel. Mooring time for the drill vessel would be reduced to connecting the vessel mooring lines to the preset explosive set anchor lines.

Drill-stem-test tools.

Mostly, drill-stem-test tools used on floating drilling vessels are those developed years ago for land operations. Until the recent activity with floaters, drill-stem tests had virtually disappeared; hence, no new tools had been developed.

In very recent years, some new tools designed especially for use on floating vessels have begun to appear. These include the subsea test tree and the slip-joint safety valve currently recommended on all drill-stem tests from floating vessels (see Chapter 16). Sperry-Sun has developed a new bottom-hole-pressure recorder that offers greatly improved sensitivity over conventional ones.

One of the most interesting and promising recent tools is an annulus-pressure-operated safety sampler developed by Halliburton[3]. Basically, the sampler is a sliding-sleeve valve in which annulus hydraulic pressure against a chamber of nitrogen opens the tool. Annulus pressure must be maintained to keep the tool open. A reduction of this pressure causes the sampler to close. Another design feature provides that if excessive annulus pressure should develop while the sampler is open the tool locks closed.

The advantage of a valve that can be selectively opened and closed by annulus pressure is the complete elimination of all vertical and rotational manipulations of the test string. In addition, the tool incorporates the features of a safety valve; that is, if the annulus loses pressure or gains it, the valve automatically closes.

This tool and similar ones that eliminate mechanical manipulation of the test string are the ultimate answer for maximum safety and reliability in drill-stem tests from a floating vessel.

Standardization of subsea equipment.

Returning a drill vessel to a previously drilled well can cost the operator a lot in modification of the equipment on the vessel. Reason is the incompatibility of the subsea equipment of various manufacturers and various rigs. A subcommittee of the American Petroleum Institute has been making an effort to solve this problem.

The incompatibility is actually limited to only two features of the subsea equipment: (1) the external configuration of the subsea wellhead; and (2) the guidepost diameter and spacing. Standardization of these items will mean that any drill vessel and its existing subsea blowout-preventer stack will be able to move onto and connect to any previously drilled well, so long as the stack and the wellhead are of the same size. The benefits to the industry will be similar to those gained in past years by the standardization of threads and flanges.

Subsea completions

There has been much work for the past ten years in the development of reliable subsea completion and production systems. All of the prototypes tested to date have been designed to be installed and maintained, at least partially, with the assistance of divers. Currently, a number of manufacturers and producers are actively working on designs for diverless, deepwater installations.

The new system that will probably be the first diverless one installed is the Humble SPS system being developed for the Santa Barbara Channel off the coast of California.[4,5] The system is based on the clustering of wells on template structures either mounted on the ocean floor or on an underwater platform. The additional cost of directionally drilling wells from a central location will be offset by the use of multiple rigs on a single floating vessel. The cost of subsea gathering systems and maintenance will also be reduced.

The template structures are being designed on a single-well modular basis, with the module engineered in detail. Ideally, a single template could provide sufficient wells for full field development. Present designs envision a system capable of handling up to 40 wells. Each template on initial installation would incorporate a manifold of all the necessary piping to handle the well from initial completion to final abandonment.

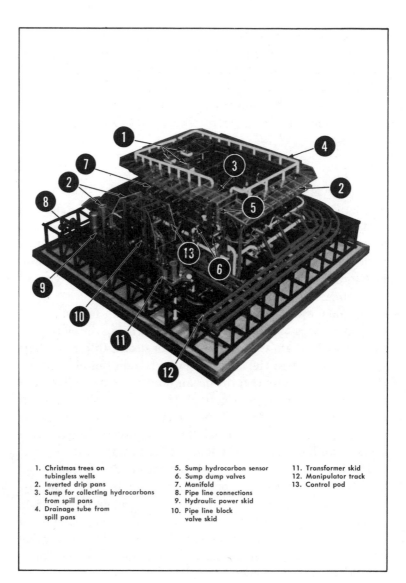

1. Christmas trees on tubingless wells
2. Inverted drip pans
3. Sump for collecting hydrocarbons from spill pans
4. Drainage tube from spill pans
5. Sump hydrocarbon sensor
6. Sump dump valves
7. Manifold
8. Pipe line connections
9. Hydraulic power skid
10. Pipe line block valve skid
11. Transformer skid
12. Manipulator track
13. Control pod

FIG. 19.3   Model of Humble's Submerged Production System

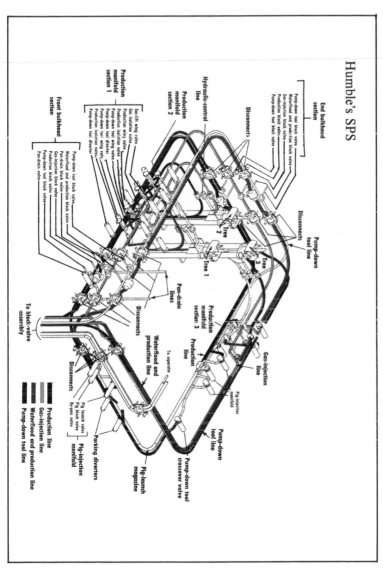

FIG. 19.4 Flow Diagram of Humble's Submerged Production System

207

It incorporates a system that permits: pump-down tools; pigging of lines; individual well tests; injection of chemicals for corrosion and hydrate control; artificial lift; and remote control of producing rates. Overall, the system proposed by Humble appears to be a feasible and efficient plan with full cognizance of the necessity for a completely safe, reliable installation.

REFERENCES

1. Butler, B., and Larralde, E.: "Motion Compensation on Drilling Vessels". Third Offshore Technology Conference, OTC-1335, April 1971.

2. Smith, J. E.: "Explosive Anchor for Salvage Operations–Progress and Status". Third Offshore Technology Conference, OTC-1504, April 1971.

3. Wray, G. Q., Petty, G. E., and Jeffords, C. M.: "Developments in Testing from Floating Vessels". SPE-AIME, SPE-3094, October 1970.

4. Scott, R. W.: "How Humble Plans to Produce Oil in the Santa Barbara Channel", World Oil, December 1970, Vol. 171, No. 7, p. 39-53.

5. Bleakley, W. B.: "Humble's Subsea Production System Nears Test Phase", The Oil and Gas Journal, August 30, 1971, Vol. 69, No. 35, p. 49-54.

Suggested Reference Material on Floating Drilling Operations and Related Subjects

1. "International Maritime Dictionary", Rene de Kerchove, D. Van Nostrand, Inc.

2. American Bureau of Shipping publications: "Rules for Building and Classing Steel Vessels;" "Rules for Building and Classing Steel Vessels for Service on Rivers and Intracoastal Waterways;" "Rules for Building and Classing Steel Barges for Offshore Service;" "Rules for Building and Classing Offshore Mobile Drilling Units;" "Guide for the Classification of Manned Submersibles;" "Surveyor", Quarterly Publication, Free of Charge.

3. Merchant-vessel publications of the U.S. Coast Guard: "Rules and Regulations for Tank Vessels" (CG-123); "Load Line Regulations" (CG-176); "Laws Governing Marine Inspection" (CG-227); "Rules and Regulations for Passenger Vessels" (CG-256); "Rules and Regulations for Cargo and Miscellaneous Vessels" (CG-257); "Rules and Regulations for Uninspected Vessels" (CG-258); "Rules and Regulations for Manning Vessels" (CG-268); "Rules and Regulations for Artificial Islands and Fixed Structures on the Outer Continental Shelf" (CG-320); and "Proceedings of the Marine Safety Council" Monthly, Free of Charge.

4. The Society of Naval Architects and Marine Engineers Publications: "Marine Engineering", Vol. 1 and 2, Herbert Lee Seward; "Ship Design and Construction", Amelio M. D'Arcangelo; "Principles of Naval Architecture", John P. Comstock.

5. "Oceanographical Engineering", Robert L. Wiegel, Prentice-Hall, Inc., 1964.

6. "Fundamentals of Construction and Stability of Naval Ships", Thomas G. Gillmer, United States Naval Institute, 2nd Edition, 1959.

7. "Ocean Sciences", E. J. Long, United States Naval Institute, 1964.

8. "Modern Seamanship", Austin M. Knight, D. Van Nostrand Co.

9. "Buoyancy and Stability of Ships", R. F. Scheltema de Here and A. R. Bakker, H. Stam Nederland N. V., Culemborg, The Netherlands.

10. Proceedings of the Joint Conference on Prevention and Control of Oil Spills, API-EPA-USCG, Washington, D.C., June 1971.

11. Preprints of the Offshore Technology Conferences, 1969, 1970, 1971, Houston, Texas (Two Volumes Each Year).

12. 1971 Directory of Marine Drilling Rigs, Published by Ocean Industry Magazine, September 1971.

Appendix A
PRELIMINARY PLANNING AND RIG EVALUATION
CHECK LIST

PRELIMINARY PLANNING AND RIG EVALUATION CHECKLIST

PART I: BASIC OPERATIONAL DATA

COMPANY_____PROSPECT/FIELD_____

ENVIRONMENTAL CONDITIONS:

Water Depth_____Tide Variation _____Current_____
Max. Waves _____Percent Occurrence_____Period_____
Other Waves_____Percent Occurrence_____Period_____
Other Waves_____Percent Occurrence_____Period_____
Max. Winds _____Percent Occurrence_____
Other Winds _____Percent Occurrence_____
Other Winds _____Percent Occurrence_____

Local Marine Activities_____

Remarks_____
_____
_____

LOCATION:

Coordinates_____
Description_____
Nearest Population Centers_____Distance_____
                          _____Distance_____
                          _____Distance_____
Dock Facilities _____Type_____Distance_____
               _____Type_____Distance_____
               _____Type_____Distance_____
Airport Facilities   _____Distance_____
Alternate Airport   _____Distance_____
Office Facilities   _____Distance_____
Warehouse Facilities   _____Distance_____
Housing Facilities   _____Distance_____
Hospital Facilities   _____Distance_____
Alternate Hospital   _____Distance_____
Oil Field Services   _____Distance_____

Remarks_____
_____
_____
_____

WELL PROGRAM:
    Estimated Spud Date_____No. of Wells_____
    Average Days/Well_____Days/Total Program_____
    Estimated Completion Date_____
    Total Depth_____Max. Formation Pressure_____

Casing Program			Mud Program	
Depth	Hole Size	Csg. Size	Depth	Mud Wt.
____	_____	_____	____	_____
____	_____	_____	____	_____
____	_____	_____	____	_____
____	_____	_____	____	_____
____	_____	_____	____	_____

Remarks_____
_____

                PART I: Prepared by_____
                         Date_____

PART II: DRILLING VESSEL EVALUATION
(Complete on each vessel considered)

Contractor_____Rig_____Type Rig_____
Displacement_____Allowable Deckload_____
Length_____Beam_____ Drilling Draft_____
Classification_____Date of Construction_____

Remarks_____

MOORING:

Anchors (1) No._____Size_____Type_____Condition_____
(2) No._____Size_____Type_____Condition_____

Chain (1) Amt._____Size_____Type_____Age_____
Condition_____Last Inspection_____
Date_____Test Load_____
(2) Amt._____Size_____Type_____Age_____
Condition_____Last Inspection_____
Date_____Test Load_____
Remarks_____

Wireline No._____Length_____Size_____Type_____
Age_____Condition_____
Dates Last Socketed_____
Remarks_____

Winches No._____Type_____Rating_____
Power_____Wire Size_____Capacity_____

Tensiometers Mfg._____Type_____Max. Load_____
Controls Location_____
Readout at_____
Printout at_____
Last Calibration_____
Frequency of Calibration_____

Mooring Pattern_____

Mooring Design	Bow	Beam
Wind, Knots	_____	_____
Wave, Feet	_____	_____
Current, Knots	_____	_____
Calc. Force on Vessel, KIPS	_____	_____
Calc. Force on Riser, KIPS	_____	_____
Calc. Total Force, KIPS	_____	_____
Tension Per Line, KIPS	_____	
Percent Breaking Strength	_____	
Initial Line Tension, KIPS	_____	

(MOORING CONTINUED)

    Max. Horizontal Displacement, %   _____

   Other Equipment

     Buoys_____

     Buoy Lights/Reflectors_____

     Pendants, No. _____Size_____Length_____

   Dynamic Positioning Equipment_____

   _____

   Remarks_____

   _____

VESSEL:

   Storage Capacities

     Dry Bulk Storage _____

     Dry Sack Storage_____

     Drilling Water_____Fresh Water_____

     Fuel Oil_____Water Distillation Units_____

   Deck Cranes (1) Mfgr._____Type_____

              Load Capacity_____Boom_____

         (2) Mfgr._____Type_____

              Load Capacity_____Boom_____

         (3) Mfgr._____Type_____

              Load Capacity_____Boom_____

         (4) Mfgr._____Type_____

              Load Capacity_____Boom_____

   Quarters A/C_____No. Men_____Recreation Room_____

        Galley_____Hospital_____

        Office Accommodations_____

   Helicopter Decks_____

   Logging Unit_____

   Cementing Unit_____

   Position Monitoring Equipment, Mfgr._____Type_____

   Pollution Control Equipment_____

   _____

   Communications Equipment_____

   _____

   Navigation Equipment_____

   Subsea Television Equipment_____

   Instrumentation_____

   Other Equipment and Remarks_____

   _____

AUXILIARY VESSELS

_____
_____
_____
_____
_____
_____

SHORE FACILITIES

_____
_____
_____
_____
_____

DRILLING EQUIPMENT:

Depth Rating of Rig_____Basis_____
Hoisting Equipment
    Mast (or derrick) Mfgr._____Size_____
        Load Capacity_____Racking Cap._____
        Last Inspection_____By_____
    Drawworks, Mfgr._____Model_____
        Power_____
_____

Crown Block, Mfgr._____Model_____Capacity_____
Traveling Block, Mfgr._____Model_____Capacity_____
Block Guidance, Mfgr._____Model_____
    Type_____
Drilling Hook, Mfgr._____Model_____Capacity_____
Drill Pipe Elevators
    (1) Size_____Mfgr._____Model_____Capacity_____
        Last Inspection_____By_____
    (2) Size_____Mfgr._____Model_____Capacity_____
        Last Inspection_____By_____
    (3) Size_____Mfgr._____Model_____Capacity_____
        Last Inspection_____By_____
Elevator Links
    (1) Size_____Mfgr._____Capacity_____
        Last Inspection_____By_____
    (2) Size_____Mfgr._____Capacity_____
        Last Inspection_____By_____
Drilling Line, Size & Type_____

(DRILLING EQUIPMENT CONTINUED)
    Remarks

_____
_____
_____
_____

Rotating Equipment
    Rotary Table, Mfgr._____Model_____Size_____
        Power_____
    Swivel, Mfgr._____Model_____Rated Load_____
    Kelly (1) Mfgr._____Size & Type_____
        (2) Mfgr._____Size & Type_____
        (3) Mfgr._____Size & Type_____
    Kelly Cock (Above Kelly)
        Mfgr._____Size_____Rating_____
    Kelly Safety Valve (Below Kelly)
        Mfgr._____Size_____Rating_____
    Spare Safety Valve (On Rig Floor)
        Mfgr._____Size_____Rating_____

Instrumentation (List Drilling Recorders, etc. Installed)

_____
_____
_____
_____
_____
_____
_____
_____

Circulating System
    Mud Pumps   Mfgr.   Model   HP Rating   Avail. Liners   Power
        1.   _____   _____   _____   _____   _____
        2.   _____   _____   _____   _____   _____
        3.   _____   _____   _____   _____   _____
        4.   _____   _____   _____   _____   _____
        5.   _____   _____   _____   _____   _____
    Shale Shakers
        1. Mfgr._____Model_____
        2. Mfgr._____Model_____
    Rotary Hose, Size_____Length_____Rating_____

(DRILLING EQUIPMENT CONTINUED)

Mud Storage Tanks	Use	Dimensions	Volume
1.	_____	_____	_____
2.	_____	_____	_____
3.	_____	_____	_____
4.	_____	_____	_____
5.	_____	_____	_____
6. Trips	_____	_____	_____

Mud Mixing Equipment_____

Pit Level Indicators_____

Flow Indicators_____

Desanders_____

Desilters_____

Centrifuges_____

Degassers_____

Remarks_____
_____

Power Plants
_____
_____
_____

Well Control Manifold & System
Manifold Size_____Rating_____
Chokes (1) Type_____Rating_____
      (2) Type_____Rating_____
      (3) Type_____Rating_____
      (4) Type_____Rating_____
Separator, Size _____Rating_____

Remarks
_____
_____

TUBULAR GOODS
  Drill Pipe (Hardbanded Drill Pipe Not Acceptable)
    (1) Amt._____Size & Wt._____Grade_____Range_____
         Tool Joints_____Condition_____
    (2) Amt._____Size & Wt._____Grade_____Range_____
         Tool Joints_____Condition_____
    (3) Amt._____Size & Wt._____Grade_____Range_____
         Tool Joints_____Condition_____
    (4) Amt._____Size & Wt._____Grade_____Range_____
         Tool Joints_____Condition_____
    (5) Amt._____Size & Wt._____Grade_____Range_____
         Tool Joints_____Condition_____
    (6) Amt._____Size & Wt._____Grade_____Range_____
         Tool Joints_____Condition_____

  Drill Collars
    (1) Amt._____Size OD_____Size ID_____Range_____
         Connections_____Condition_____
    (2) Amt._____Size OD_____Size ID_____Range_____
         Connections_____Condition_____
    (3) Amt._____Size OD_____Size ID_____Range_____
         Connections_____Condition_____
    (4) Amt._____Size OD_____Size ID_____Range_____
         Connections_____Condition_____
    (5) Amt._____Size OD_____Size ID_____Range_____
         Connections_____Condition_____
    (6) Amt._____Size OD_____Size ID_____Range_____
         Connections_____Condition_____

  Remarks

  _____
  _____
  _____
  _____
  _____
  _____
  _____
  _____
  _____

MARINE RISER, TENSIONING, GUIDANCE SYSTEMS:
  Marine Riser Size OD_____ID_____ Grade Material _____
            Mfgr. & Type Connectors_____
            Amount_____No. of Joints_____

(MARINE RISER, TENSIONING, GUIDANCE SYSTEMS CONTINUED)

List Pup Joints_____

Last Inspection_____By_____

Frequency of Inspections_____

Kill & Choke Lines, How Installed_____

_____Size_____

Telescoping Joint, Mfgr._____Size (ID)_____

Length of Stroke_____

Diverter System, Describe_____

Flexible Joint, Mfgr._____Model or Type_____

Max. Deflection_____Angle Indicator?_____

Riser Tensioning System_____

Maximum Tension_____Angle Line with Riser_____

Last Calibration Tension Gauges_____Frequency_____

Riser Tensioning Lines, Size_____Description_____

Guideline Tensioning System_____

_____Maximum Tension_____

Last Calibration Tension Gauges_____Frequency_____

Guideline Size_____Description_____

Remarks_____

_____

_____

BLOWOUT PREVENTER STACK:

Blowout Preventers (Main Stack)

(1) Mfgr. & Type_____Size_____Rating_____

(2) Mfgr. & Type_____Size_____Rating_____

(3) Mfgr. & Type_____Size_____Rating_____

(4) Mfgr. & Type_____Size_____Rating_____

(5) Mfgr. & Type_____Size_____Rating_____

(6) Mfgr. & Type_____Size_____Rating_____

Available Pipe Rams_____

Locking Devices on Preventer Nos._____

Ram Position Indicators on Preventer Nos._____

Kill & Choke Valves (Main Stack)

(1) Mfgr. & Type_____Size_____Rating_____

Failsafe?_____Hyd. Opened?_____Hyd. Closed?_____

(2) Mfgr. & Type_____Size_____Rating_____

Failsafe?_____Hyd. Opened?_____Hyd. Closed?_____

(BLOWOUT PREVENTER STACK CONTINUED)

    (3) Mfgr. & Type_____Size_____Rating_____

       Failsafe?_____Hyd. Opened?_____Hyd. Closed?_____

    (4) Mfgr. & Type_____Size_____Rating_____

       Failsafe?_____Hyd. Opened?_____Hyd. Closed?_____

Side Outlet Locations (1)_____(2)_____

Accumulators (Mounted on Stack)

    No._____ Mfgr._____Type_____

    Charge Pressure_____Rated Volume_____Usable Vol._____

Automatic or Acoustical Closing Device?_____

                Which Preventer?_____

Top Connector, Mfgr. & Type_____Size_____Rating_____

    Last Inspection_____Frequency_____

Bottom Connector, Mfgr. & Type_____Size_____Rating_____

    Last Inspection_____Frequency_____

Auxiliary Stack (Describe) _____

_____

_____

Remarks_____

_____

_____

CONTROL SYSTEM:

Manufacturer_____Type System (Indirect) (Direct)

Control Pods (If indirect system) Mfgr. & Type_____

Surface Accumulators

    No._____Mfgr._____Type_____

    Charge Press._____Rated Volume_____Usable Volume_____

    Calc. Volume Open & Close All Preventers_____

       Operating Press._____Surplus Volume %_____

      (Include subsea accumulators in calculations)

Control Hoses

    Hose Bundle No. 1 Length_____Reel Capacity_____

    Hose Mfgr. & Type_____

Hose No.	Size	Function	Hose No.	Size	Function
1	____	_____	21	____	_____
2	____	_____	22	____	_____
3	____	_____	23	____	_____
4	____	_____	24	____	_____
5	____	_____	25	____	_____
6	____	_____	26	____	_____

(CONTROL SYSTEM CONTINUED)

Hose No.	Size	Function	Hose No.	Size	Function
7	___	_____	27	___	_____
8	___	_____	28	___	_____
9	___	_____	29	___	_____
10	___	_____	30	___	_____
11	___	_____	31	___	_____
12	___	_____	32	___	_____
13	___	_____	33	___	_____
14	___	_____	34	___	_____
15	___	_____	35	___	_____
16	___	_____	36	___	_____
17	___	_____	37	___	_____
18	___	_____	38	___	_____
19	___	_____	39	___	_____
20	___	_____	40	___	_____

Closing Time: Hydril-type Preventer_____

Ram-type Preventers _____

Hose Bundle No. 2 Length_____Reel Capacity_____

Hose Mfgr. & Type_____

Hose No.	Size	Function	Hose No.	Size	Function
1	___	_____	21	___	_____
2	___	_____	22	___	_____
3	___	_____	23	___	_____
4	___	_____	24	___	_____
5	___	_____	25	___	_____
6	___	_____	26	___	_____
7	___	_____	27	___	_____
8	___	_____	28	___	_____
9	___	_____	29	___	_____
10	___	_____	30	___	_____
11	___	_____	31	___	_____
12	___	_____	32	___	_____
13	___	_____	33	___	_____
14	___	_____	34	___	_____
15	___	_____	35	___	_____
16	___	_____	36	___	_____
17	___	_____	37	___	_____
18	___	_____	38	___	_____
19	___	_____	39	___	_____
20	___	_____	40	___	_____

(CONTROL SYSTEM CONTINUED)

    Closing: Hydril-type Preventer_____

                Ram-type Preventers_____

    Remarks_____

_____

GENERAL INFORMATION:

    Overall Appearance of Vessel and Equipment_____

_____

    Number and Experience Level of Contract Personnel_____

_____

    Past Operating Performance of Vessel_____

_____

    Past Safety Record of Vessel and Personnel_____

_____

    Past Company Experience With Contractor_____

_____

    Remarks_____

_____

Source of Above Data_____

PART II Prepared by_____

           Date_____

Appendix B

U. S. COAST GUARD SUGGESTED SELF-INSPECTION
CHECK-OFF LIST FOR MOBILE DRILLING
UNITS OPERATING OVERSEAS
(PRELIMINARY COPY)

SAFETY INSPECTION FOR
U.S. COAST GUARD

Name of Vessel: _____ O.N. _____

Date of this Inspection: _____

Date of last Inspection: _____

Location of Vessel: _____

Name of Owners: _____

Address of Owners: _____

Type of Propulsion: _____

Type of Hull: _____

Name of Master: _____ Lic. No. _____

Name of Chief Engineer _____ Lic. No. _____

Name of Company Safety Inspector: _____

In my opinion the vessel (does) (does not) meet all the rules and regulations applicable to this class vessel and is in a safe and seaworthy condition.

_____
(Signature & Title)

                                                  YES   NO   DATE OF
                                                              EXPIRATION

1. Certificate of Inspection
2. Amendments to the Certificate of
       Inspection
3. Safety Construction Certificate
4. Safety Equipment Certificate
5. Exemption Certificate
6. Radiotelegraphy Certificate
7. Radiotelephony Certificate
       Other Certificates to be on Board

                                                  YES   NO

1. Vessel Document
2. Loadline Certificate properly endorsed
       Placards Posted

                                                  YES   NO

1. Instruction for launching inflatable
       liferafts
2. Station Bills
3. Instructions for fire extinguishing
       systems
4. Stability Letter
5. General arrangement plans showing fire
       stations
6. Atomic Attack Instructions (CG-3256)
7. Lifesaving signals & Breeches Buoy
       Instruction, CG-811
8. Fire & Boat Drill (CG-809)

VESSEL MARKINGS (97.40)

YES   NO

1. Loadline—legible—conforms to
    certificate
2. Draft marks legible—proper size—
    properly spaced
3. Vessel Name & Hailing Port
4. Net Tonnage
5. Official Number

PUBLICATIONS

YES   NO

1. Laws governing Marine Inspection
2. Rules & Regulations for Class
3. Loadline Regulations
4. Pertinent Navigation & Vessel
    Inspection Circulars
5. Trim & Stability Booklet

DIRECTIONS

The following safety check-off list is designed to include the items of inspection which would be covered by a Coast Guard inspector at an inspection for certification, re-inspection and drydocking. The list was prepared to cover self-propelled vessels, barges and column stabilized units. Each inspector should mark those items which are applicable to the particular vessel under inspection.

It is realized that many drilling units will be classed by a recognized classification society; however, these surveyors do not inspect the safety devices such as firefighting equipment and lifesaving devices. Therefore, much of the emphasis of this report has been placed on these safety items.

A check-off list for the complete hull structure would be lengthy and difficult to include all the various hull scantlings, components and arrangements. Each safety inspector should have sufficient knowledge of the hull design and aware-ness of the strength requirements to give those areas and parts the painstaking examination required to detect any deficiencies which will need attention. When deficiencies are noted, repair or procurement action should be initiated with follow up action to determine that the requirement has been satisfied.

This safety check-off list is divided into the following parts as an orderly sequence for inspection and reporting:

A. Lifesaving Equipment
B. Fire Protection
C. Hull Structure and Fittings
D. Navigational Equipment
E. Machinery
F. Electrical
G. Drydocking

The following publications and circulars should be available for reference:
A. Rules and Regulations for Cargo and Miscellaneous Vessels, CG-257
B. Marine Engineering Regulations, CG-115
C. Electrical Engineering Regulations, CG-259
D. Navigation and Vessel Inspection Circular Numbers:
    (1) 2-63, Guide for Inspection and Repair of Lifesaving Equipment
    (2) 14-65, Guide to Firefighting Equipment aboard Merchant Vessels
    (3) 7-68, Notes on Inspection and Repair of Steel Hulls
    (4) 12-69, Special Examination in Lieu of Drydocking for large Mobile Drilling Units

A. LIFESAVING EQUIPMENT

I. *Discussion:*
Lifesaving equipment is on a vessel solely for the purpose of serving as a safety function. These items are not used from day to day, but are required to perform in an emergency. It is essential that they be of good quality, suitable for the purpose intended, that they be maintained in good condition, and that they be ready for immediate use in the event of an emergency. Lifesaving equipment may appear to be in good condition, but only thorough examination and testing will assure the actual condition. Emergency equipment should not be taken for granted, because in time of need, it must be there and working and the crew must be acquainted with its operation. With this in mind, the following inspection should be conducted. It is recommended that Navigation and Vessel Inspection Circular No. 2-63, entitled "Guide for Inspection and Repair of Lifesaving Equipment" be reviewed prior to conducting the inspection. The law requires that any life jacket found on board to be so defective as to be incapable of repair, shall be destroyed.

II.   *Lifeboat Installation*

	1	2	3	4
Lifeboat Capacity (number of persons)				
Each lifeboat is Coast Guard approved				
Stripped, cleaned & overhauled (date)				
Hull examined for defects				
Hull material				
Checked equipment-replaced out dated items				
Boat releasing gear operates freely				
Suspension Test-weight added (date)				
Operated & examined winches and davits				
Checked boatfalls for fishhooks & defects				
Tested boat engine				
Examined fuel tank				
Date fuel replaced				
Steering apparatus turns freely				
Tested air tanks				
All markings legible				
Radio installation				
Checked limit switches & electric controls				
Date boat falls last renewed				

III.   Rigid Lifefloats. Examined straps, webbing waterlight, launching device
and painter:_____
Number on board:_____

IV.   Inflatable Liferafts: Serial No.   _____
Capacity   _____
Serviced by authorized agent
    (date)   _____
Name of servicing agent   _____
Stowage satisfactory with
    container drain holes face
    downward   _____
Test of hydrostatic release
    (properly marked)   _____

V.   Examined rescue boat hull & fittings: Condition_____
Hull material:_____
Date stripped and cleaned: _____
Stowage & launching facilities: _____

VI.   Number of life preservers on board: _____
All life preservers examined: _____ were
found to be defective and _____ replaced.
All life preservers marked with vessels name. YES_____NO_____
Required marking legible on storage lockers. YES_____NO_____
Number of work vests: _____. All were examined and _____
were found defective and _____ were replaced.

VII.   *Ring life buoys:*
Total No. on board   _____

    _____
No. found defective   _____
No. with waterlights   _____
No. of waterlights defective   _____
No. with 15 fathom line   _____
No. with self-activating
    smoke signals   _____
All properly marked   YES_____NO_____

VIII.   Line throwing apparatus: (Required on all mechanically propelled
vessels of 150 gross tons and over in ocean or coastwise service).
The equipment and appliances were examined and found to be: _____
Date tested:_____

IX.   *Distress signals* (vessel over 150 gross tons).
Stowed on the pilothouse or navigation bridge.

Required 12 approved hand-held rocket-propelled parachute red flare distress signals, contained in a portable container. Note: the service limited to a period of three years from date of manufacture.
Number on Board:_____Date of manufacture: _____

X.   *Portable radio apparatus* (vessel on international voyage)
   (a) Meets requirements of FCC: _____
   (b) Radio is stowed:_____
   (c) Date last tested: _____
Remarks:

XI.   The master is responsible for conducting a fire and boat drill at least once in every week. Each drill shall be conducted as if an actual emergency existed. All hands should report to their respective stations and be prepared to perform the duties specified in the station bill.
   (a) The vessel's log indicated that the required drills have been conducted. YES_____NO_____
   (b) A fire and boat drill was witnessed by the inspector on _____ ___and was *(un)satisfactory.*
If unsatisfactory, list reasons.

B. FIRE PROTECTION

I.   *Discussion:*
Fire protection equipment is on a vessel solely for the purpose of serving as a safety function. These items are not used from day to day, but are required to perform in an emergency. It is the duty of the master, or person in charge to see that the vessel's fire-fighting equipment is at all times ready for use and that all such equipment required by the regulations is provided, maintained, tested, and replaced as necessary. Designated fire hose should not be utilized for vessel wash down or for other ship cleanup procedures. It is recommended that Navigation and Vessel Inspection Circular No. 14-65, entitled "Guide to Fixed Fire-Fighting Equipment Aboard Merchant Vessels", be reviewed prior to conducting the inspection. The law requires that any fire hose found that is so defective as to be incapable of repair, shall be destroyed.

II.   *Firemain System:*
Number of fire stations:_____
All stations have been examined with full fire main pressure. The piping, cutoff valves, drains, and hydrant were satisfactory. All stations had the required nozzles, spanners, fog nozzles, applicators, strainers, and markings. YES_____NO_____
Remarks:
The overall condition of the fire main is: _____
Every compartment can be reached by two different fire hoses, one from a single length. YES_____NO_____
Remarks:

III.   *Fire hose:*
Vessel is required to carry_____ft. of fire hose. All fire hose was of the approved type and was tested and found satisfactory, or the defective hose was renewed. YES_____NO_____
Remarks:

IV.   The fire pump(s) was tested and relief valve was found to relieve at_____psi. The discharge pressure gage was in_____
_____condition. The vessel has an International Shore Connection. YES_____NO_____
Remarks:

V. *Fire detection system:* (If applicable)
The vessel is equipped with a_____fire detection system. The entire system was examined and tested and found to be _____ .

VI. *Fixed system:*

The vessel is fitted with _____fixed system(s).
                          (number)
They protect the following spaces.

1._____   4. _____
2._____   5. _____
3._____   6. _____

The equipment has been examined and the alarms, controls, instruction and markings are satisfactory, except_____
_____

CO_2 bottles weighted (date) _____
If serviced by a private company, list name _____
CO_2 bottles hydro-tested (date)_____
Foam containers (refilled)_____.

Deck foam systems shall be tested biennially by discharging foam for approximately 15 seconds from any nozzle. It is not necessary to deliver foam from all foam outlets, but all lines and nozzles shall be tested with water to prove them to be clear of obstruction. Each foam system utilizing a mechanical foam system shall submit a representative sample of the foam liquid to the manufacturer who will issue a certificate indicating gravity, pH, percentage of water dilution, and solid content.

Date of last test _____ .
Date of last certificate _____ .
    Gravity_____
    Solid content _____
    pH_____
    % water dilution_____

VII. *Hand portables:*
Number of fire extinguishers: *Required Onboard Spares*
                        A-II
                        B-II
                        C-II

All fire extinguishers have been examined in accordance with 46 CFR, Table 91.25-20 (a) (1) and found to be in satisfactory condition, and tagged. YES_____ NO___ _____.
Remarks: _____

_____

There are spare charges on board for at least 50% of each size and variety of hand portable fire extinguishers. YES_____NO_____

NOTE: If the unit is of such variety that it cannot be readily charged by the vessel's personnel, one spare unit of the same classification shall be carried in lieu of spare charges for all such units of the same size and variety.

VIII. *Semi-portable fire extinguishers:*

Location	Type		Condition of		Date tested, weighted & tagged
		Controls	Hose	Nozzles	

IX. *Semi-portable pressurized dry chemical extinguishers:*
Visual inspection indicates:
    a. sufficient quantity of dry chemical (date) _____
    b. there is a free flow of dry chemical—YES_____NO_____
    c. tested remote release (date)_____
    d. pressure vessel examined (date) _____

X. *Fire Axes;*
Number required_____. Number on board_____
All axes were examined and the blades were sufficiently sharp, the handles secure enough to withstand the repeated impact of the blade against wood or thin metal.   YES_____ NO_____.

XI. *Fireman's Outfit:*
Number on board _____ .
each self-contained breathing apparatus with lifeline, explosion-proof flashlight, and flame safety lamp was examined and found to be in
_____ condition.
There was one complete recharge for each outfit and spare battery for each flashlight.
The equipment is located_____and_____number of the crew are familiar with the operation of the equipment.

XII.   *Sand:*
         On vessels of over 1000 gross tons, containing an oil-fired boiler, there shall be a metal receptacle containing not less than 10 cubic feet of sand, sawdust impregnated with soda, or other approved dry material together with a scoop or shaker. On vessels of 1000 gross tons or less, at least 5 cubic feet of such material shall be carried. Or in lieu of the above, on B-II fire extinguisher may be substituted. The following is on board:_____

XIII.   If the vessel has a helicopter facility, then the fire-fighting capability must be examined to determine that it is maintained in accordance with the approved plans.
         Describe Vessel's Installation:

C. HULL STRUCTURE AND FITTINGS

I.   *Discussion:*
The hull structure must be thoroughly examined to insure that hull strength is intact and that watertight integrity is not impared. Although each drilling unit may be designed differently, they all will be built with structural features necessary to withstand stresses and strains of its own design. Column stabilized units should be examined using the same philosophy for a normal displacement hull, but considering the strength and design aspects of the columns and tabular structure, with a working platform or deck connected at the top. Bracing members used to connect the columns or lower hull and the truss system supporting the working platform must also be included in the strength consideration. Keeping this in mind, the inspection and examination of the entire structure must be accomplished. Defects, deterioration and damage of the main strength members and the watertight envelope must be detected and corrected. It is recommended that Navigation and Vessel Inspection Circular No. 7-68, entitled "Notes on Inspection and Repair of Steel Hulls", be reviewed prior to conducting the inspection.

II.   After inspection of the hull structure (hull girder and superstructure), list: (1) the compartments which were inaccessible and (2) any deterioration or structural defects found:

III.   During this examination, the inspector should also see that the spaces are reasonably clean and that a prolonged accumulation of old line, canvass, loose gear, valves, drums, drilling equipment, and other miscellaneous stores has not created a fire hazard. The stowage of paint and oil in an unapproved locker is strictly prohibited. The bilges in all spaces should be clean and free of fire hazards.
Remarks:

IV   While making the rounds of the vessel, items such as ladders, stairways, guardrails, storm rails and other common structural accessories are examined to be determined that they are in good condition.

V.   Some compartments will have reach rods to bilge drainage valves. These should be exercised to determine that they rotated freely. Bilge piping should be checked for soundness.
Remarks:

VI.   Bilge drainage should be checked. The system should be operated to determine that suction is possible. The rose boxes and strainer

plates are clear, clean and sound. Bilges are free of fire hazards.
Remarks:

VII.   It is not unusual to find defective light fixtures and uncapped outlets in seldom used spaces. The condition of the electrical wiring and fixtures should be checked to determine the safe condition.
Remarks:

VIII.   All penetration through watertight bulkheads should be checked for tightness.
Remarks:

IX.   The paint and lamp lockers should be checked for proper construction, stowage, missing vapor globes and guards, and correct installation and markings of the fire protection equipment.
Remarks:

X.   Deck openings and closures should be checked:
    (a)   for soundness, and that the closing devices operate fully and the gaskets are in good condition.
          Remarks:
    (b)   Number of watertight doors in subdivision bulkheads _____ .
All tested and found to operate satisfactorily, are tight and in good condition.   YES_____NO_____
    (c)   Number of side ports_____ .
All examined and found to be tight and in good condition.   YES_____NO_____
    (d)   Number of main deck hatches and scuttles_____ .
All examined and found to be tight and in good condition.   YES_____NO_____

XI.   The anchor chains, devil claws, pelican hooks, chain stoppers, riding pawls, hawsepipe and hawsepipe covers were checked and found in _____
_____ condition.
The anchor windlasses were operated (if possible). The brakes and hoisting machinery were found to be _____ .

XII.   All other mooring equipment and machinery was examined and found to be in _____ condition.

XIII.   The condition of all other mooring devices, bits, bulwarks, cleats, pad eyes, fairleads, and plain or roller-type chocks were examined and found to be in _____ condition.
Remarks:

XIV.   The condition of the vessel's cargo handling gear such as winches, booms, goosenecks, blocks, runners, pins, shackles and other fittings have been examined and found to be in _____ condition.
Remarks:

XV.   The vessel arrangement provides two avenues of escape from every general area within the vessel where crew and workmen may be quartered or normally employed.
YES_____NO_____
NOTE: It is not always possible to provide such exits from the holds or columns, the means of escape provided should be adequately safe and easily accessible.

XVI.   *Ventilation:*
(a) Adequate for all compartments (including hazardous areas).
YES_____NO_____
(b) Remote controls to power ventilation marked and tested.
YES_____NO_____
(c) Closures for spaces protected by fixed smothering systems examined and satisfactory.   YES_____NO_____
(d) All fuel tank vents examined and found sound. Screen in place and in good condition.   YES_____NO_____

XVII.   Sanitary inspection of the quarters, heads, galley, reefer boxes and dry store room revealed_____ deficiencies. The lighting was adequate, and any safety hazard removed. The galley exhaust vent was checked and found to be free of grease and litter.
YES_____NO_____
Weekly sanitation inspections have been conducted by master and chief engineer.   YES_____NO_____

XIX.   The hospital space was inspected and found to be clean and properly equipped with medical supplies.
There are_____bunks.
Remarks:

XX.   This vessel is required to have structural fire protection.
YES_____NO_____
The vessel meets the requirements and the approved material is in satisfactory condition.   YES_____NO_____

XXI.   If the vessel is fitted with a helicopter facility, then the structure should be examined for deterioration of structural defects.
Remarks:

D. NAVIGATIONAL EQUIPMENT

I.   *Discussion:*
      All items of importance located in the vicinity of the pilothouse having
      to do with the vessel's navigation should be examined and tested.
      It is realized that many drilling vessels will remain on one station for
      long periods between moves. The navigational equipment should be
      thoroughly checked out well in advance of any planned move to insure
      that the equipment will be functioning properly. Some of these items
      such as the steering system, the telephone and telegraph system and
      pertinent electrical systems require collaboration with the deck and
      engineering personnel. All voiced tubes should be tested for audibility,
      along with the other electrical and sound powered telephone systems.
      Upon testing the steering system and telegraph systems, the inspector
      must be satisfied that the equipment is functioning freely and accurately
      and that the various alternate methods and alarms are in proper working
      order. The vessel should have all the necessary navigational instru-
      ments, charts and publications. Each vessel should have the proper
      navigational lights, whistle, fog horns, fog bells and on vessels over
      350 ft., a fog gong, and distress signals. It is recommended that all
      the electronic equipment such as radar, loran, gyrocompasses, gyro
      repeaters, gyro pilot, radio direction finder, depth sounder, position
      indicator, etc., be connected to the emergency power system if prac-
      tical, and be maintained in good condition. Other items such as hand
      lead, flag signals, jury rigged breakdown and towing lights should be
      checked for condition and adequacy.

II.   During the inspection of the navigational lights, the following should
      be noted:
      (a)   The screens on the port and starboard running lights are
            painted glossy black.   YES_____NO_____
      (b)   The portable cable leading from the light fixture to a permanent
            outlet is in good condition, not unduly long, and not in the
            nature of a jury rig.   YES_____NO_____
      (c)   The navigational light panel was examined by testing each
            light switch in all positions and checking all fuses for proper
            size type   YES_____NO_____
      (d)   Lights found with one filament of a double filament bulb burned
            out have been replaced.   YES_____NO_____
      (e)   The navigational panel alarm was checked.
                                                YES_____NO_____
      Remarks:

III.   Navigation equipment was examined and tested as follows:

	Condition	Remarks
Radio Direction Finder		
Radio		
Whistle		
Sounding Equipment		
Fog Bell & Gong		
Radar		
Engine Order Telegraph		
Rudder Angle Indicator		
Flag Signals		
Navigational Light		
Navigational Shapes		
Vessel Telephone System		
Magnetic Compasses		
Gyro Compass & Repeaters		
Loran		
Pilot Ladder		

E. MACHINERY

I.   *Discussion:*
       At each examination, the inspector shall ensure that the main and
       auxiliary machinery, boilers and their appurtenances, pressure vessels

and equipment are in satisfactory operating condition and fit for the service for which they are intended. Special attention shall be given to correcting personnel hazards. Such things as gears, couplings, flywheels and all machinery capable of injuring personnel shall be provided with adequate covers or guards and all protective devices must be adequately maintained.

II.   (a)   The propulsion machinery is classed by_____.

(b)   The machinery is in good operating condition, the foundations are sound, the guards are in place and the controls function properly. All safety devices and speed trips have been checked and found to be operating satisfactorily.

                                             YES_____NO_____

Remarks:

III.   Boilers (propulsion and auxiliary)
        Vessel is equipped with following boiler(s):
          1.
          2.

IV.   Boilers are built in a variety of arrangements, styles, and sizes with accessories, controls, and operating features. Regardless of their design, failure of a pressure part could result in a serious casualty with probable loss of life. Each boiler must be cooled, cleaned, examined, inspected, and tested in accordance with good marine engineering practices. The exterior of each boiler should be examined with close attention to the foundation. The fireside should be cleaned and the refractory repaired as necessary. The watersides should be inspected for sign of corrosion and pitting. The general condition of the tube ends, internal fittings and their supports should be examined. The boilers should be hydrostatically tested as required by the regulations. The mounting should be opened for inspection every four years and removed every eight years. The safety valves should be checked to determine the relieving pressure and percent blowdown. The hand relieving gear on the safety valve should be checked to determine that it operates freely.
        (a)   Condition of boiler exterior:
        (b)   Condition of firesides:
        (c)   Condition of watersides:
        (d)   Date of last hydrostatic tests:
        (e)   Date mountings last opened:
        (f)   Date mountings last removed:
        (g)   Date safety valves last tested or relieved:

(h) . Has it been necessary to break the Coast Guard seal on any safety valve? YES_____NO_____
If YES, describe the facts. Was this reported to the U.S. Coast Guard? If so, where?

(i) Tested the boilers under full operating pressure and found these boilers to be in safe condition.
.   YES_____NO_____

Remarks:

V.   Inspected steam piping. Found the lagging or insulation, hangers and supports in _____ condition. All steam piping subject to boiler pressure, was last given a hydrostatic test of _____ psi on _____ (Date)

VI.   The fuel system was examined under operating pressure and the remote control fuel shut-off valves were examined and tested to insure that the fuel pump can be remotely shut down in the event of a fire or casualty.
      YES_____NO_____

VII. The feed water system (including the condenser) was examined under operating conditions and found to be in _____ condition.

VIII. Unfired Pressure Vessels:

Service	Working Pressure	Relief Valve Setting	Date Tested or Examined

IX.   *Steering Gear:*
      The steering gear was tested and examined. The examination included all devices employed in the change-over from automatic to manual operation, the rudder control and follow up devices, the emergency steering wheel and trick wheel. The condition of this equipment is:___

X.   *Auxiliary Machinery:*
      (1) All auxiliary machinery has been examined and found to be in safe operating conditions.
           YES_____NO_____
      (2) Remote controls for the means of stopping machinery driving forced and induced draft fans, fuel oil transfer pumps, fuel oil unit pumps and fans in the ventilation system serving machinery have been tested and found to be in working condition.
           YES_____NO_____

XI.   *Heating Boilers:*
      Each heating unit has been carefully examined and an operational

test conducted to insure that all the control components and safety components are in satisfactory operating condition.

YES_____NO_____

Date of last hydrostatic test: _____

XII. *Miscellaneous Safety Valves and Relief Valves:*
The inspector shall observe that the safety valves or relief valves installed on reduced pressure lines, evaporators, feedwater heaters, air receivers, etc., operate satisfactorily. The safety and relief valves shall be tested or relieved to determine that they can prevent the building-up of an excess pressure before the shut-off valve can be closed and serve as a warning in the event of the failure of the reducing valve. The setting of such valves has been checked and found satisfactory or adjusted if found necessary.

YES_____NO_____

XIII. *Bilge Pumps:*
The vessel has_____bilge pumps.
The operation of the pumps has been observed and found to be satisfactory.

YES_____NO_____

The bilge piping, eductors, injectors, manifolds, valves, strainers, sounding and vent pipes have been inspected and found to be in_____

_____ condition.

XIV. *Refrigeration and Air Conditioning System:*
The compressors, controls, valves, and alarms have been examined and found to be in_____

condition. The operation of the equipment is satisfactory.

YES_____NO_____

Refrigeration space alarm tested.   YES_____NO_____

Refrigeration mask was examined and found to be in _____ condition.
(If ammonia system is installed, the mask is not required)

XV. The compressed air system was examined and found to be in _____

_____ condition.

XVI. The sanitary system was examined and found to be in _____ condition.

XVII. The fresh water (portable & domestic) system was examined and found to be in _____

_____ condition.

XVIII. The evaporator was examined and found to be in_____ condition.
The safety valves were tested and found to be satisfactory.

YES_____NO_____

F. ELECTRICAL

I.   *Discussion:*
      The inspector shall examine the electrical installation to insure that
      the arrangements and materials comply with the approved drawings
      and applicable rules and regulations. The primary purpose of the
      electrical inspection is to assure an adequate, reliable, and safe elec-
      trical system for marine service, the components of which provide
      safety to personnel from electrical shock and minimize the danger
      of fire originating within the electrical system. Particular attention shall
      be placed to guards which will prevent personnel from coming into
      contact with dangerous electric currents.

II.   The vessel is equipped with_____electrical generators. Each unit
      was examined and found to be in_____condition.
      The overspeed trips were checked and found to operate satisfactorily.
                                               YES_____NO_____
      All protective guards were in place.
                                               YES_____NO_____
      Remarks:

III. *All* switchboards were examined and found to be in_____
      condition. The protective guards and insulating matting were in place,
      the back of the board was not jury wired to by-pass protection devices,
      the fuses were of the proper type and size and the switchboard is
      protected against falling objects and dripping water.
                                               YES_____NO_____
      The reverse current or reverse power relays have been tested and
      are satisfactory.
                                               YES_____NO_____
      Remarks:

IV.   The various motors on board the vessel have been examined and
      found to be in_____condition. The cables and leads are
      in good condition or have been replaced. Protrusions on rotating parts
      such as couplings are covered by guards.
                                               YES_____NO_____
      Remarks:

V.   The various controllers have been examined. Watertight units were
      examined for deterioration, holes, and faulty gaskets. Drip-proof enclo-
      sures are adequately protected against falling objects and moisture.
      A legible wiring diagram is posted on the inside of the enclosure door.
      No jury rig or jumper circuits exist. YES_____ NO_____.
      Remarks:

VI.   Interior Communication Systems such as the soundpowered tele-
phones, voice tubes, engine order telegraph and rudder angle indicator
have been examined, tested and found to be satisfactory.

YES_____NO_____

Remarks:

VII.   General alarm was examined. The condition of the batteries and wiring
was_____. The fuses were of the proper size and the
system was properly marked.

YES_____NO_____

The system was tested and each alarm bell operated properly.

YES_____NO_____

VIII.   The emergency generator was examined and tested satisfactorily under
the normal emergency load for a reasonable period of time.

YES_____NO_____

Remarks:

IX.   Batteries are located in a safe space protected from falling objects.
The vent system was checked and operates properly; the explosion-
proof lighting fixtures have their globes and guards intact.

YES_____NO_____

X.   (a)   The lighting systems have been examined. The fixtures were
checked for globes and guards in place where required. The
distribution panels were checked for the correct size fuses, and
correct directories. The cable was checked for general condition
and jury rig wiring eliminated. Lighting fixtures were checked for
correct wattage of lamps.

YES_____NO_____

(b)   The emergency lighting system was checked and found to be
in satisfactory condition and all lamps provided adequate illumina-
tion.

YES_____NO_____

(c)   All emergency lights are legibly marked with the letter "E" at
least ½ inch high.

YES_____NO_____

Remarks:

XI.   A check of all hazardous areas was made to determine that all electrical
equipment located therein was suitable for the location, that it was
in good condition and had not been altered, and that it was properly
assembled and maintained.

YES_____NO_____

Remarks:

G. DRYDOCKING

I. *Discussion:*

(a) The regulations require drydocking and tailshaft examinations for all inspected vessels. It is unlikely that a Coast Guard inspector will be in attendance at overseas locations; however, in most instances a classification surveyor will be on hand. The Classification Society's report may be submitted to supplement the inspection report; nevertheless, the master or company inspector should complete this form as evidence of his evaluation of the underwater body examination. The Navigation and Vessel Inspection Circular No. 7-68, entitled "Notes on Inspection and Repair of Steel Hulls", should be reviewed at the time of the drydocking. For those large mobile drilling units which are so large that they cannot be handled by any existing drydock facility, a special examination may be provided as outlined in Navigation and Vessel Inspection Circular No. 12-69.

(b) The inspector should examine all shell plating for indents, corrosion, grooving and other defects. Where excessive wastage is suspected, the plate must be tested for thickness. An examination should be made of all welding and rivets. Particular attention must be directed to piping, fittings, sea chests, spool pieces and sea valves. The stern frame, rudder, rudder post, pintles, gudgeons, skeg, propellers, tailshaft and tailshaft bearings should be thoroughly inspected. Clearances of the tailshaft bearing should be taken and the tailshaft drawn for inspection on appropriate years as required by the regulations. Frequently the rudder post gland is neglected during the drydocking and later leakage has flooded the hull spaces. The vessel should be gas free in those areas where repair work will be accomplished.

II. The exterior of the vessel's shell plating has been thoroughly examined and the general condition and finding are as follows:

III. All sea chest strainers were removed. The sea valves were opened and examined. The sea chests were thoroughly examined. All were found in_____

_____ condition.

IV. The rudder and the associated appurtenance and bearings were examined, were found to be in _____ _____ condition.

V. The propeller was examined and found to be in _____ _____ condition.

VI. (a) The tailshaft weardown was taken and found to be _____ .
(b) The tailshaft was drawn and examined and found to be_____ .

VII. The anchor chains were ranged and found to be in _____ condition.

VIII. The following repairs were accomplished to hull and structure: (List the Coast Guard office where plan approval was accomplished.)

IX. Outstanding repair to be accomplished at a later drydocking or repair period.

Appendix C

THEORY FOR CALCULATION OF WAVE FORCES

From American Bureau of Shipping: Rules
for Building and Classing Offshore Mobile
Drilling Units, 1968.

251

PART 1

SHALLOW WATER WAVE THEORY

The method presented is a simplification based on an interpolation between the solitary and Airy theories, and several others. The analysis is based on vertical cylindrical structures and thus may be used for units having structural and stability columns or, without serious error, truss type legs with non-cylindrical components. The method also assumes that the structure extends to the bottom of the sea. In the event that the legs or columns stop short of the bottom, it may either be assumed that the forces have diminished greatly at such point, and the non-existent portion below ignored, or an adjustment may be made, finding the effective wave height at that distance below the water, and making another calculation of the imaginary portion below the actual structure, and subtracting from the original value.

Formulas

$$F_{Dm} = \tfrac{1}{2} \times C_D \times \rho \times D \times h^2_w \times K_{Dm}$$
$$F_{im} = \tfrac{1}{2} \times C_m \times \rho \times D^2 \times h_w \times K_{im}$$
$$L_w = 5.12T^2 \times \frac{L_w}{L_a} \times \frac{L_a}{L_o}$$
$$M_{Dm} = S_D \times F_{Dm}$$
$$M_{im} = S_i \times F_{im}$$
$$M_{Tm} = \frac{F_m}{F_{Dm}} \times M_{Dm}$$

$L_a =$ linear theory wavelength for period T and depth h, feet
$L_o =$ deepwater linear theory wave length $= 5.12T^2$, feet
$h =$ still-water depth, feet
$T =$ wave period, seconds
$\zeta_o =$ crest elevation above still water, feet
$F_{Dm} =$ maximum value of total horizontal drag force, pounds (occurs at wave crest)
$C_D =$ drag coefficient (use 0.71 for following example)
$W =$ unit weight of sea water, pounds/cubic feet
$g =$ acceleration of gravity, 32.2 feet/sec^2
$$\rho = \text{mass density} = \frac{W}{g} = \frac{64}{32.2}$$
$$= 1.993 \text{ slugs/cubic feet}$$
$D =$ pile diameter, feet
$h_w =$ wave height, crest to trough, feet
$K_{Dm} =$ drag force factor at crest, feet/sec^2

F_{im} = maximum value of total horizontal inertial force, lb (occurs at between crest and ¼ of wavelength)

S_D = lever arm for F_{Dm}, feet

S_i = lever arm for F_{im}, feet

C_m = inertial or mass coefficient (use 2.00 for following example)

K_{im} = inertial force factor, feet/sec^2

M_{Dm} = moment on pile about bottom associated with maximum drag force, foot pounds.

L_w = wave length, including the effect of finite wave height, feet

F_m = maximum value of combined drag and inertia forces, pounds

S = lever arm for F_m, feet

M_{im} = moment on pile about bottom associated with maximum inertial force, foot pounds.

M_{Tm} = maximum total moment on pile about bottom, foot pounds

β = angular position of maximum moment ahead of wave crest, degrees.

Example

Given: Wave height h_w = 35 feet
       Still-water depth h = 85 feet
       Wave period T = 12 seconds
       Pile diameter D = 8 feet
       Drag coefficient C_D = 0.71
       Inertia coefficient C_m = 2.00

Compute: $\dfrac{h_w}{T^2} = \dfrac{35}{144} = 0.243$ feet/sec^2 . . .   (1)

$\dfrac{h}{T^2} = \dfrac{85}{144} = 0.590$ feet/sec^2 . . .   (2)

$\dfrac{h_w}{h} = \dfrac{35}{85} = 0.412$ . . . .   (3)

From Figure 1 using Eqs. 1 and 2, get $\dfrac{\zeta_o}{h_w} = 0.68$

$\zeta_o = 0.68 \times h_w = 0.68 \times 35 = 23.8$ feet

From Figure 2 using Eq. 2, get $\dfrac{L_a}{L_o} = 0.75$

From Figure 3, using Eqs. 1 and 2, get $\dfrac{L_w}{L_a} = 1.04$

$L_w = 5.12 \times T^2 \times \dfrac{L_a}{L_o} \times \dfrac{L_w}{L_a} = 5.12 \times 144 \times 0.75 \times 1.04 = 575$ feet

From Figure 4, using Eqs. 1 and 2, get $K_{Dm} = 13.0$ feet/sec^2

$F_{Dm} = \frac{1}{2} \times 0.71 \times 1.993 \times 8 \times 35^2 \times 13.0 = 90{,}200$ lb

From Figure 5, using Eqs. 2 and 3, get $\dfrac{S_D}{h} = 0.91$

$S_D = 0.91 \times h = 0.91 \times 85 = 77.4$ feet

$M_{Dm} = F_{Dm} \times S_D = 90,200 \times 77.4$
$\quad\quad = 6,980,000$ ft-lb

From Figure 6, using Eq. 2, get $K_{im} = 19.5$ feet/sec^2

$F_{im} = \tfrac{1}{2} \times 2.00 \times 1.993 \times 8^2 \times 35 \times 19.5 = 87,200$ lb

From Figure 7, using Eqs. 2 and 3, get $\dfrac{S_I}{h} = 0.78$

$S_I = 0.78 \times h = 0.78 \times 85 = 66.3$

$M_{Im} = F_{im} \times S_I = 87,200 \times 66.3$
$\quad\quad = 5,780,000$ ft-lb

$\dfrac{F_{im}}{F_{Dm}} = \dfrac{87,200}{90,200} = 0.967$

From Figure 8, using $\dfrac{F_{im}}{F_{Dm}} = 0.967$, get $\dfrac{F_m}{F_{Dm}} = 1.37$

$F_m = 1.37 \times F_{Dm} = 1.37 \times 90,200$
$\quad\quad = 123,500$ lb

$M_{Tm} = 1.37 \times M_{Dm} = 1.37 \times 6,980,000$
$\quad\quad\quad = 9,560,000$ ft-lb

Maximum total force: $\quad\quad\quad\quad F_m = 123,500$ lb.

Maximum total moment:

$\quad M_{Tm} = 9,560,000$ ft-lb

$\quad$ Lever $= \dfrac{M_{Tm}}{F_m}$

$\quad\quad S = 77.4$ feet

From Figure 9 position of maximum moment ahead of wave crest:

$\dfrac{D^2 h}{h_w^2 L_w} = \dfrac{8^2 \times 85}{35^2 \times 575} = 0.00772, \ \beta = 13$ degrees

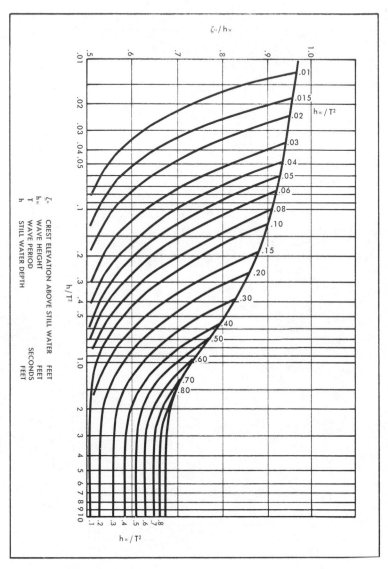

FIG. C-1 Ratio of Crest Elevation to Wave Height

256

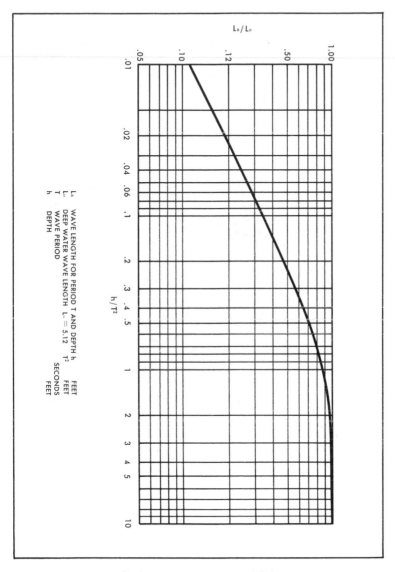

FIG. C-2 Relative Wave Height

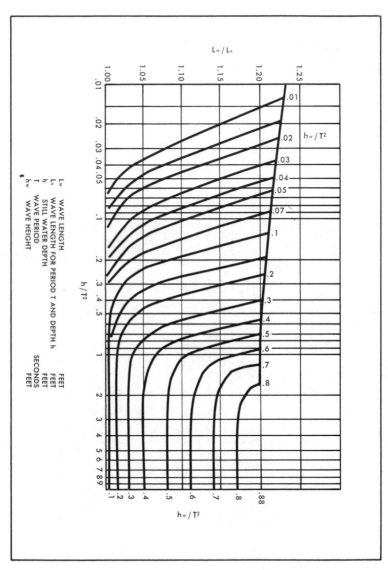

FIG. C-3 Wave Length Correction Factor for Steepness

258

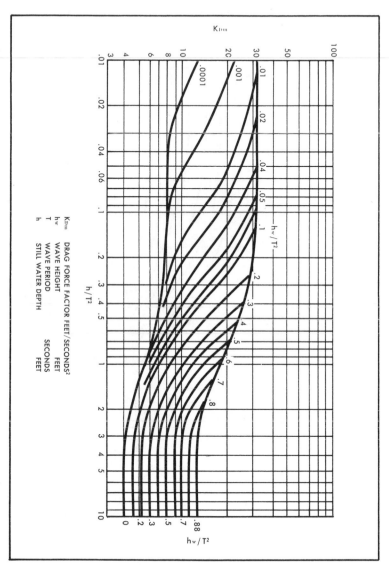

FIG. C-4 Drag Force Factor

259

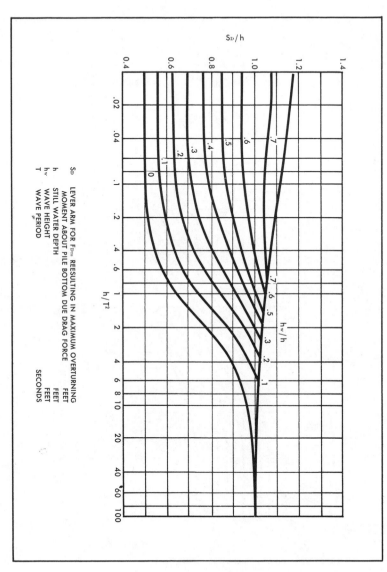

FIG. C-5 Drag Force Lever/Still Water Depth

260

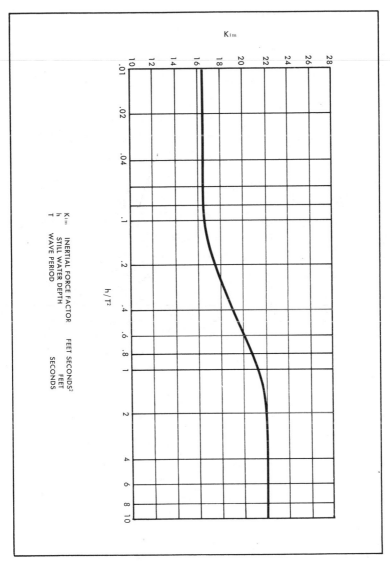

FIG. C-6 Inertial Force Factor

261

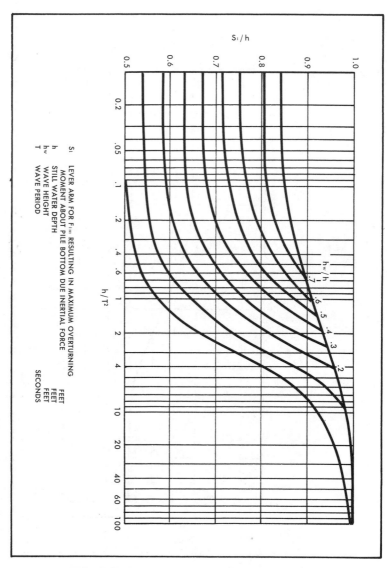

FIG. C-7 Inertial Force Lever/Still Water Depth

262

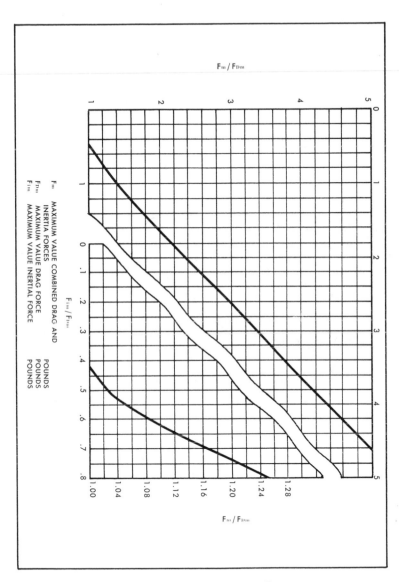

FIG. C-8 Total Force / Drag Force

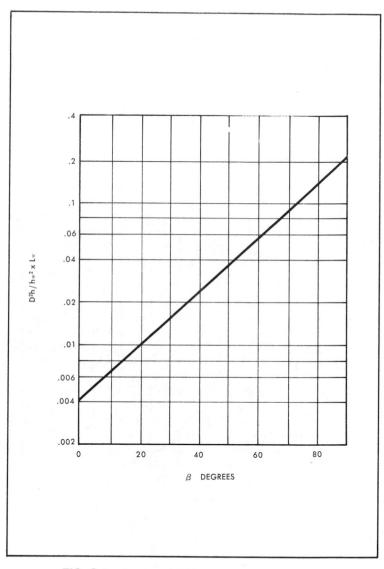

FIG. C-9 Angular Position of Maximum Moment

264

PART 2
WAVE THEORY FOR DEEP WATER

This part of Appendix C is a development of the sine wave theory for deep water waves and may be used for determining the drag and inertial forces on the underwater portions of drilling units which may be operating in location where the depth of water exceeds three hundred feet. Other methods of determining the force which may be deemed appropriate will be considered provided they are referenced and supported by calculations.

SURFACE WAVE EQUATION

$$Z = \frac{H}{2} \cos(kx - \omega t)$$

Where:

$$k = \frac{2\pi}{\lambda} \qquad \lambda = \text{wave length, feet}$$

$$\omega = \frac{2\pi}{T} \qquad T = \text{wave period, seconds}$$

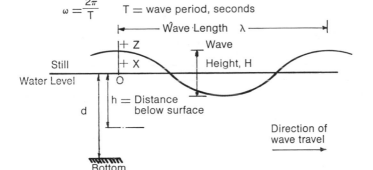

At Point "h" Below Surface, "x" from Origin

Formula $d > \frac{1}{2} \lambda$

Water Velocity:

Horizontal, $V_x = \dfrac{\omega H\, e^{-kh}}{2} \cos(kx - \omega t)$

Vertical, $V_z = \dfrac{\omega H\, e^{-kh}}{2} \sin(kx - \omega t)$

Water Acceleration:

Horizontal, $\alpha_x = \dfrac{\omega^2 H\, e^{-kh}}{2} \sin(kx - \omega t)$

Vertical, $\alpha_z = \dfrac{-\omega^2 H\, e^{-kh}}{2}\ \cos\,(kx-\omega t)$

Dynamic Pressure $P = (\rho g)\,\dfrac{H}{2}\ e^{-kh}\cos\,(kx-\omega t)$

ρ = mass density of water = 1.99 lb secs²/ft⁴

g = gravity = 32.16 ft/sec²   (ρg) = 64 lb/cft

$\lambda = gt^2/2\pi$

1. The *total* pressure at any point at a distance h below the surface is the static pressure = $\rho g h$ plus the wave dynamic pressure given above.

Surface

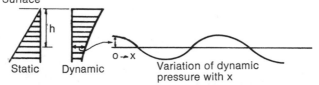

Static   Dynamic   Variation of dynamic
pressure with x

2. Note that the slope of the dynamic pressure diagram is equal to the water acceleration.

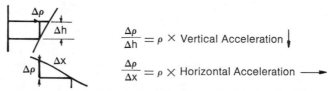

$\dfrac{\Delta p}{\Delta h} = \rho \times$ Vertical Acceleration $\downarrow$

$\dfrac{\Delta p}{\Delta x} = \rho \times$ Horizontal Acceleration $\longrightarrow$

Thus, for a narrow body, in the direction of flow, accelerations may be used instead of differences in pressure, to determine inertia forces.

EXAMPLE OF DETERMINING INERTIA FORCE
IN DEEP WATER

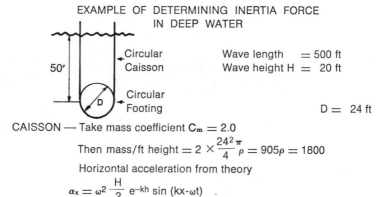

50′   Circular
Caisson

Circular
Footing

Wave length   = 500 ft
Wave height H = 20 ft

D = 24 ft

CAISSON — Take mass coefficient C_m = 2.0

Then mass/ft height = $2 \times \dfrac{24^2}{4}\pi\,\rho = 905\rho = 1800$

Horizontal acceleration from theory

$\alpha_x = \omega^2\,\dfrac{H}{2}\ e^{-kh}\sin\,(kx-\omega t)$.

$$\omega^2 = \frac{(2\pi)^2}{T} = \frac{2\pi g}{\lambda} = \frac{202}{500} = 0.40 \quad k = \frac{2\pi}{\lambda} = 0.0125$$

Then, the force per foot, at a point h below surface,

$$F_h = m x \alpha_x = 1800 \times \frac{0.4 \times 20}{2} \, e^{-.0125h} \sin (kx - \omega t)$$

$$= 7200 \, e^{-.0125h} \sin(kx - \omega t)$$

$$\left(\begin{array}{l} \text{The total force and its center may be determined by} \\ \text{calculating several values for h between 0 and 50 ft} \\ \text{and using Simpson's rule, or by integrating as follows:} \end{array} \right)$$

Total force on caisson

$$F_c = 7200 \sin (kx - \omega t) \int_0^{50} e^{-.0125h} \, dh$$

$$= \frac{7200}{.0125} (1 - e^{-.625}) = 267{,}500 \sin (kx - \omega t) \text{ lb}$$

Moment of force from surface

$$M_c = 7200 \sin (kx - \omega t) \int_0^{50} h e^{-.0125h} \, dh$$

$$= \frac{7200}{(.0125)^2} (1 - 1.625 \, e^{-.625}) = 5{,}980{,}000 \sin (kx - \omega t) \text{ ft-lbs}$$

FOOTING — with same mass/ft as caisson,
and h = 50 ft
Force per foot of length

$$F_f = 7200 \, e^{(-.0125 \times 50)} = 3850 \sin (kx - \omega t) \text{ lbs/ft}$$

DRAG FORCE IN DEEP WATER

Where appropriate the drag forces are calculated in a manner similar to the inertia forces as shown in Appendix C (Part 2) using the velocity equations shown in Appendix C (Part 2), and drag coefficients as listed in Section 3 of these Rules.

RECOMMENDED MASS FACTORS

I. TWO-DIMENSIONAL VALUES OF C_m

Condition	Shape	C_m
Submerged	Circular	$\begin{cases} 2.0 \text{ large diam.} \\ 1.5 \text{ small diam.} \end{cases}$
Submerged	Ellipse	$1.0 + b/h$
Submerged	Flat Plate	1.0
		(with cylinder area; $\pi b^2/4$)
Submerged	Rectangle	$1.0 + b/h$
Floating	Rectangle	$1.0 + b/2h$ (vertical)
Floating	Rectangle	$1.0 + b/2h$ (horizontal)
On-Bottom	Rectangle	$1.0 + 2b/h$ (horizontal)

II. THREE-DIMENSIONAL CORRECTION TO C_m

For all shapes, multiply C_m by factor, K

$$K = \frac{(l/b)^2}{1 + (l/b)^2}$$

III. APPLICATION

Immersed

Total mass = (volume of body) $\times$ K $\times$ C_m $\times$ ρ

ρ is mass density of water = $\dfrac{\text{unit weight}}{\text{gravity}}$ and is

equivalent to 2 in engineering units.

IV. NOMENCLATURE

h is dimension parallel to flow
b is section breadth normal to flow
l is length of body (normal to flow)
(lb is plane normal to flow)

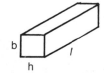

Flow

V. CORRELATION

See Page 266 for check with theory.

Index